Effect of IL-10 and anti-TGF-beta
antibodies on the morphology of
bone marrow stroma cultures.
from
Interleukin-10
by
Jan E. DeVries and
René de Waal Malefyt
© RG Landes Co. 1995

MOLECULAR
BIOLOGY
INTELLIGENCE
UNIT

NITRIC OXIDE:
A MODULATOR
OF CELL-CELL INTERACTIONS
IN THE MICROCIRCULATION

Paul Kubes, Ph.D.
University of Calgary
Calgary, Alberta, Canada

R.G. LANDES COMPANY
AUSTIN

MOLECULAR BIOLOGY INTELLIGENCE UNIT

NITRIC OXIDE: A MODULATOR OF CELL-CELL INTERACTIONS IN THE MICROCIRCULATION

R.G. LANDES COMPANY
Austin, Texas, U.S.A.

Please address all inquiries to the Publisher:
R.G. Landes Company, 909 Pine Street, Georgetown, Texas, U.S.A. 78626
or
P.O. Box 4858, Austin, Texas, U.S.A. 78765
Phone: 512/ 863 7762; FAX: 512/ 863 0081

U.S. and Canada ISBN 1-57059-280-2

International ISBN 3-540-60036-1

While the authors, editors and publisher believe that drug selection and dosage and the specifications and usage of equipment and devices, as set forth in this book, are in accord with current recommendations and practice at the time of publication, they make no warranty, expressed or implied, with respect to material described in this book. In view of the ongoing research, equipment development, changes in governmental regulations and the rapid accumulation of information relating to the biomedical sciences, the reader is urged to carefully review and evaluate the information provided herein.

Library of Congress Cataloging-in-Publication Data

Nitric oxide : a modulator of cell-cell interactions in the microcirculation /
    [edited by] Paul Kubes.
        p. cm. — (Molecular biology intelligence unit)
    Includes bibliographical references and index.
    ISBN 1-57059-280-2 (alk. paper)
    1. Nitric oxide—Physiological effect. 2. Microcirculation. 3. Cell interaction.
    I. Kubes, Paul. II. Series.
        [DNLM: 1. Endothelium-Derived Relaxing Factor. 2. Nitric Oxide.
        3. Microcirculation—physiology. 4. Cell Communication. QV 150 N7313 1995]
QP106.6.N58 1995
612.1'35—dc20
DNLM/DLC   95-14126
for Library of Congress   CIP

Publisher's Note

R.G. Landes Company publishes five book series: *Medical Intelligence Unit, Molecular Biology Intelligence Unit, Neuroscience Intelligence Unit, Tissue Engineering Intelligence Unit* and *Biotechnology Intelligence Unit*. The authors of our books are acknowledged leaders in their fields and the topics are unique. Almost without exception, no other similar books exist on these topics.

Our goal is to publish books in important and rapidly changing areas of medicine for sophisticated researchers and clinicians. To achieve this goal, we have accelerated our publishing program to conform to the fast pace in which information grows in biomedical science. Most of our books are published within 90 to 120 days of receipt of the manuscript. We would like to thank our readers for their continuing interest and welcome any comments or suggestions they may have for future books.

Deborah Muir Molsberry
Publications Director
R.G. Landes Company

CONTENTS

EDITOR

Paul Kubes
Immunology Research Group
University of Calgary
Calgary, Alberta, Canada
Chapters 2, 4, 5

CONTRIBUTORS

Joseph S. Beckman
Departments of Anesthesiology
  and Biochemistry
The University of Alabama
  at Birmingham
Birmingham, Alabama, U.S.A.
Chapter 1

D. Neil Granger
Department of Physiology
  and Biophysics
Louisiana State University
  Medical Center
Shreveport, Louisiana, U.S.A.
Chapter 6

Dwight D. Henninger
Department of Physiology
  and Biophysics
Lousiana State University
  Medical Center
Shreveport, Louisiana, U.S.A.
Chapter 6

Iwao Kurose
Department of Physiology
  and Biophysics
Louisiana State University
  Medical Center
Shreveport, Louisiana, U.S.A.
Chapter 6

Lianxi Liao
Department of Physiology
  and Biophysics
Louisiana State University
  Medical Center
Shreveport, Louisiana, U.S.A.
Chapter 6

Marek W. Radomski
Departments of Obstetrics and
  Gynaecology and Pharmacology
Perinatal Research Centre
University of Alberta
Edmonton, Alberta, Canada
Chapter 3

Brian K. Reuter
Intestinal Disease Research Unit
Faculty of Medicine
University of Calgary
Calgary, Alberta, Canada
Chapter 7

Eduardo Salas
Departments of Obstetrics and
  Gynaecology and Pharmacology
Perinatal Research Centre
University of Alberta
Edmonton, Alberta, Canada
Chapter 3

John L. Wallace
Intestinal Disease Research Unit
Faculty of Medicine
University of Calgary
Calgary, Alberta, Canada
Chapter 7

PREFACE

Inflammation is a localized response induced by microbial infection and/or cellular and tissue injury. An inability to produce an inflammatory response leads to a compromised host. On the other hand, inappropriate activation of the inflammatory cascade or more importantly absence of a "stop" signal in an otherwise normal inflammatory process leads to debilitating inflammatory diseases. There is tremendous scientific interest in the many mechanisms that may "turn-on" or maintain an inappropriate inflammatory condition. However, the mechanisms that under normal, nonpathological conditions "turn off" or counter the inflammatory signals remain essentially unknown. Understanding the endogenous mechanisms that "turn off" inflammation would certainly help in the development of new and perhaps more effective approaches to anti-inflammatory drug design. After all, the most efficient mechanism to regulate and turn off the inflammatory process is not presently prescribed therapy, but an unidentified endogenous mechanism(s) used daily by our host defense system. This book in part explores the possibility that endogenous nitric oxide may be this turn off mechanism.

The key features of the inflammatory response are leukocyte recruitment into the target tissue via the microvasculature, an increase in microvascular permeability (edema formation) followed closely by cellular and tissue injury and ultimately organ dysfunction. Major strides have been made over the past decade to understand the role of adhesion molecules used by leukocytes to infiltrate tissue and cause inflammation. Recent work has suggested that nitric oxide may be a homeostatic regulator of leukocyte and platelet infiltration as well as a regulator of moment-to-moment changes in fluid and protein movement across the vasculature. Moreover, nitric oxide may also regulate the trafficking of inflammatory cells to sites of injury and inflammation. The specific focus of this book is to examine nitric oxide as a potential regulator of the physiology and pathophysiology of the microcirculation with special emphasis on nitric oxide and cell-cell interaction. Therefore, the areas of nitric oxide biology covered in this book have leukocyte adhesion as an important underlying element. The chapters of this book can be divided into three parts. The first chapter introduces the chemistry of nitric oxide as it may behave in the microcirculation. The next three chapters of the book summarize the role of nitric oxide as it pertains to leukocyte adhesion, platelet adhesion and microvascular permeability. The next section focuses on three pathophysiologic conditions wherein leukocyte adhesion has been invoked as essential to the pathogenesis of the disease and the role that nitric oxide may play. Ischemia/reperfusion, hypercholesterolemia and atherosclerosis are highlighted in this book, and the data with respect to nitric oxide in each of these conditions are summarized. The final chapter reviews very recent work on the development of a novel series of nonsteroidal anti-inflammatory drugs modified to incorporate a nitric oxide-releasing moiety, which unlike the parent compounds, lacks gastrointestinal toxicity perhaps by reducing leukocyte adhesion.

Each of the contributors has presented the current state of knowledge and identified areas of controversy in this expanding field of science. I would like to personally thank the various investigators who devoted their valuable time in writing critical chapters for this book. They are world leaders in their respective areas of research and certainly have many other very important commitments. I would also like to acknowledge the different agencies that provide funding to allow us to continue to do essential medical research.

CHAPTER 1

BIOCHEMISTRY OF NITRIC OXIDE AND PEROXYNITRITE

Joseph S. Beckman

The discovery that mammalian cells produce nitric oxide has revolutionized our understanding of the control of blood flow, thrombogenesis, gastrointestinal motility and modulation of neuronal activity.[1,2] Nitric oxide is the smallest and structurally simplest intracellular messenger, consisting of only a nitrogen and an oxygen. Although its chemistry has been extensively studied by inorganic chemists (see Jones[3]), the applications to biological systems are fraught with misunderstandings and unnecessary complications. Most articles describe nitric oxide as highly reactive and a gas, but in biological settings it is neither. Nitric oxide reacts slowly with most biological molecules including thiols. It is also dissolved in solution like many uncharged molecules including glucose, oxygen and carbon dioxide. Nitric oxide is comparable in reactivity to molecular oxygen, which also is dissolved in cells and reacts at slow rates with most biological molecules. However, both nitric oxide and oxygen can be converted to much stronger oxidants by secondary reactions. In particular, the diffusion-limited reaction of nitric oxide with superoxide to form peroxynitrite is an important component of the microbicidal actions of macrophages and neutrophils and can also cause substantial injury in many disease states. Thus, a simple understanding of the biological chemistry of nitric oxide can help explain how such a simple molecule can be produced by so many diverse tissues with so many diverse actions.

Nitric Oxide: A Modulator of Cell-Cell Interactions in the Microcirculation, edited by Paul Kubes. © 1995 R.G. Landes Company.

Nitrogen has seven electrons and oxygen eight. Thus, there are a total of fifteen electrons in a molecule of nitric oxide. Molecular orbitals can contain a maximum of two electrons, each with opposite spins. (If this simple rule could be violated, the universe would collapse into a single orbital with a blinding flash of light.) Nitric oxide can have seven orbitals with paired electrons, but the eighth orbital will have an unpaired electron. Consequently, nitric oxide is a free radical. While we commonly think of free radicals as highly reactive and toxic, like the hydroxyl radical (HO·), many are not. For example, oxygen in the ground state is a biradical with two electrons occupying separate orbitals. Many transition metals including copper and iron have orbitals with unpaired electrons. Iron can exist in the high spin ferrous oxidation state with five unpaired electrons.

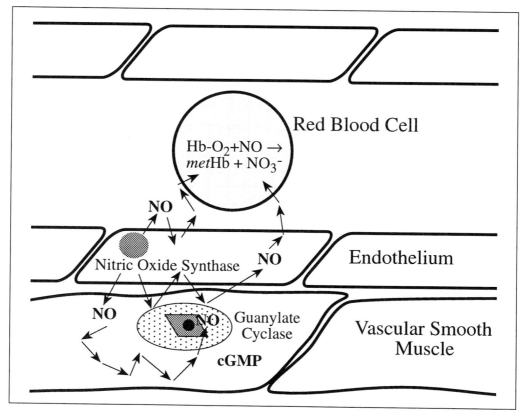

Fig. 1.1. Diffusion paths and reactions of nitric oxide in the vascular wall. Arrows illustrate the convoluted paths that nitric oxide may follow. In one second, 37% of nitric oxide would have diffused further than 55 µm, a distance nine times greater than that of a red blood cell.

Free radicals tend to react rapidly with other molecules with unpaired electrons. For example, oxygen binds tightly to the ferrous iron in hemoglobin. Nitric oxide binds even more strongly to ferrous iron in porphyrin groups.[4] The major route of signal transduction by nitric oxide is activation of soluble guanylate cyclase, a protein that contains ferrous protoporphyrin IX, like hemoglobin, with the iron in the ferrous state. This protein has a high affinity for nitric oxide, with concentrations as low as 5-10 nM being sufficient to activate the protein. Remarkably, endogenous concentrations of oxygen (~200 µM) do not combine with guanylate cyclase. It has been proposed that binding of nitric oxide to the heme in guanylate cyclase weakens the affinity to a distal histidine, which then allows activation of the enzyme. Preparations of guanylate cyclase with the porphyrin removed are active and not affected by nitric oxide.[5]

The major route for eliminating nitric oxide in vivo is almost certainly by reaction with *oxy*-hemoglobin or *oxy*-myoglobin. Only one atom of oxygen binds to the ferrous heme of hemoglobin. Because oxygen is a biradical, the second oxygen is available with an unpaired electron for reaction with nitric oxide. The second order reaction of nitric oxide with *oxy*-hemoglobin is particularly rapid (10^8 $M^{-1} \cdot s^{-1}$), only a 100-fold lower than the diffusion limit for two molecules colliding with each other. The reaction proceeds through an intermediate

$$Hb\text{-}Fe^{2+}\text{-}O\text{-}O\bullet + \bullet NO \rightarrow Hb\text{-}Fe^{2+}\text{-}OONO \rightarrow Hb\text{-}Fe^{3+} + NO_3^-$$

and results in the formation of *met*-hemoglobin and nitrate.[6] The *met*-hemoglobin is recycled by reductases, and the nitrate (NO_3^-) is secreted in the urine. The production of nitric oxide is the only known route to form nitrite and nitrate in mammals. When nitrate and nitrite intake in the diet were carefully controlled in humans, an excess of nitrite was found in the urine.[7] A huge increase occurred in one volunteer who developed the flu during the study.[7] This provided some of the first evidence for the elevated production of nitric oxide by activated inflammatory cells in humans.

How can *oxy*-hemoglobin be the major sink for nitric oxide when hemoglobin is confined in red blood cells and nitric oxide is produced in tissues? The answer is that nitric oxide is highly diffusable, with a diffusion coefficient at 37°C slightly larger than that of molecular oxygen.[8,9] The maximum distance from a blood vessel is typically 100-200 µm. Clearly, oxygen is able to diffuse the same distance into a tissue in sufficient quantities to supply mitochondria. For a small hydrophobic molecule like nitric oxide, it would take only a few seconds to diffuse through a tissue to a blood vessel. Cell membranes pose no barrier for nitric oxide, which can be crossed as easily as by oxygen or carbon dioxide. It is blood vessels that limit the diffusion of nitric oxide in tissues.

Diffusion is key for understanding the biological activity of nitric oxide. The shortest half-life of nitric oxide reported in guinea pig heart is just under a second.[10] Although a second is brief, it is still many times longer than a heart beat, a muscle contraction or a neuron firing. In a second, a sprinter can run nearly ten meters. In a second, nitric oxide may spread through several cell diameters, with 37% of the nitric oxide produced by a point source spreading further than 57 µm.[11] Consequently, diffusion away from an individual cell is far more rapid than reactions of nitric oxide with endogenous molecules within a cell. As a corollary, the local concentration of nitric oxide becomes significantly greater when a cluster of cells produce nitric oxide simultaneously, even within the cells producing nitric oxide. When you are stuck in a traffic jam, most of the nitric oxide you are breathing was produced by other cars and not by your own automobile exhaust.[12]

The rapid diffusion of nitric oxide can integrate biological responses only on a local scale, because the spread of the message is limited by the vasculature. Nitric oxide is unusual for an intracellular messenger because it cannot be distributed through the blood stream due to the rapid elimination by hemoglobin. In effect, nitric oxide behaves like a shock absorber. The time delay of neuronal networks to initiate and modulate muscle movements could easily lead to parasitic oscillations without a means to dampen feedback control of the responses. Without nitric oxide, many physiological activities will behave like an automobile on a bumpy road with damaged shock absorbers.

The biological importance of nitric oxide in controlling blood flow locally was elegantly demonstrated by Griffith et al.[13] If blood flow suddenly increases in an artery due to dilation of a downstream vascular bed, turbulent flow could result from the increased blood flow. Turbulence creates shear stress that would evoke contraction of vascular smooth muscle, which would cause a catastrophic collapse of blood flow by further increasing shear stress as the vessel contracts. Neurogenic vasodilation would be difficult to control because of the time delay to process and transmit appropriate signals could allow parasitic oscillations to develop. However, shear stress on endothelial cells causes them to release nitric oxide, which will induce the underlying smooth muscle to dilate. As the vessel dilates, shear stress is reduced, and the production of nitric oxide decreases. The local action and significant lifetime of nitric oxide relative to rapid changes in blood flow modulate the responses of a blood vessel, thereby minimizing turbulence. However, atherosclerotic lesions develop at sites subjected to frequent turbulence, which might be related to the prolonged exposure to nitric oxide and its secondary oxidants in these regions.

Nitric oxide has similar functions in the central nervous system, where its production is controlled by the entry of calcium that in turn depends upon a neuron having been particularly active. A local cluster of neurons that have been active for a fraction of a second can produce a focal accumulation of nitric oxide that will diffuse throughout the local region. Nitric oxide can affect synaptic plasticity within that region.[14] Consequently, nitric oxide can be a retrograde messenger, carrying and integrating information about the activity of local neurons to the surrounding synaptic connections. This activity will allow neuronal behavior to be modified and may contribute to the formation of memories.[15,16]

BIOSYNTHESIS OF NITRIC OXIDE

Nitric oxide is produced by the oxidation of arginine. The requirement for arginine was independently discovered by two groups. Long before nitric oxide was known to be produced in vivo, Deguchi and Yoshioka[17] purified arginine from crude brain extracts by assaying cofactors necessary for the activation of guanylate cyclase in brain. Hibbs and coworkers independently discovered that the tumoricidal activity of activated macrophages depends upon arginine being present in the media.[18] They were also the first to use the substrate-based inhibitor, N-methylarginine, which greatly decreased the tumoricidal activity of macrophages.

While nitric oxide was first discovered in endothelial cells because of its biological activity of relaxing isolated blood vessels, the first purification of the enzyme was from brain because the brain and the cerebellum in particular contain at least 20-fold greater concentrations of nitric oxide synthase than the remainder of the body.[19,20] The enzyme was discovered to be a modified form of cytochrome P450 fused with a reductase. One of the terminal nitrogens on arginine is oxidized by the enzyme in two steps, initially to produce N-hydroxyarginine[21] and then to the final products citrulline and nitric oxide. Both steps occur in the active site, which contains a heme group similar to other P450s and biopterin.[22] Although biopterin is a commonly used cofactor in the hydroxylation of phenylalanine and tyrosine, its role in the activity of nitric oxide synthase is not as clear.[23] In tyrosine hydroxylases, the biopterin is released from the enzyme to be recycled, whereas it remains tightly bound in nitric oxide synthase. Without biopterin, nitric oxide synthase becomes uncoupled and consumes many fold more oxygen per nitric oxide produced.

For each nitric oxide produced, one arginine, two oxygens and 1.5 NADPHs are consumed. One atom from the molecular oxygen ends up in the nitric oxide, one on citrulline and the other two are reduced to water. The overall reaction involves a net re-

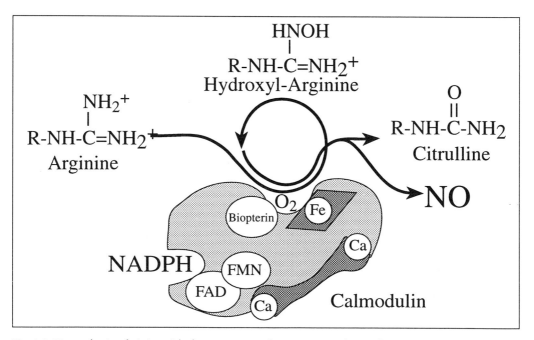

Fig. 1.2. Biosynthesis of nitric oxide from arginine. The enzyme is a homodimeric protein, though only one subunit is illustrated here. The catalytic cycle produces hydroxyarginine as an intermediate, which remains bound to the enzyme for further oxidation to citrulline and nitric oxide.

duction of five electrons. Two electrons are supplied by the substrate, arginine, and three are provided by NADPH. However, NADPH is a two electron carrier. Nitric oxide synthase also contains two flavin cofactors which are able to hold either one or two electrons. Thus, the extra electron from the second NADPH is utilized for the oxidation of a second arginine. The enzyme responsible for producing superoxide in activated macrophages and neutrophils also stores electrons from NADPH on flavins and transfers them one at a time to molecular oxygen.

Three basic classes of nitric oxide synthase have been isolated and cloned.[24] The endothelial and brain derived isozymes are expressed normally and are regulated by calmodulin and physiologically relevant concentrations of calcium. These are commonly called constitutive isoforms, though like most constitutive enzymes, their transcription and translation are carefully regulated and are altered under stress and during development. The inducible form of nitric oxide synthase was originally isolated from activated macrophages.[25] However, cytokines and bacterial lipopolysaccharides have been shown to induce nitric oxide synthase in almost every tissue examined.[26-28] The inducible form of nitric oxide synthase is not regulated by physiologically relevant concentrations of calcium and is commonly identified as the calcium-independent isoform. It came as a surprise to discover that the inducible nitric oxide synthase was tightly bound to calmodulin, which copurified with the enzyme.[29] Because calmodulin is small (12 kDa) compared to the inducible nitric oxide synthase (135 kDa), it often ran off the bottom of SDS electrophoresis gel. The affinity of calmodulin for the inducible nitric oxide synthase is so high that the enzyme remains fully active even at the lowest physiologically relevant levels of calcium.

The rate of nitric oxide production is similar for each of the three isoforms of nitric oxide synthase.[24] However, the inducible form can be induced in much higher expression and can produce enormous amounts of nitric oxide. The evolutionary advantage of virtually any cell being able to induce the rapid production of nitric oxide is not certain. It may be a defense against viral infection.[30] Nitric oxide is known to inactivate ribonucleotide reductase, a key enzyme in the biosynthesis of DNA.[31,32] Nitric oxide may also contribute to the development of apoptosis, an endogenous process of initiating cell death that may limit viral replication and the development of cancer.[33,34] The mechanisms of cytotoxicity mediated by nitric oxide are relatively unknown and a source of considerable confusion in the literature at present.

NITRIC OXIDE TOXICITY

The basis for nitric oxide reactivity in biological settings is considerably different than what might be inferred from the well characterized reactions of nitric oxide reported in inorganic chemistry textbooks. As a result, nitric oxide is assumed to directly attack many molecules like DNA or thiols, whereas the reaction rates are very slow and probably not of great biological significance.[35]

If one mixes pure nitric oxide gas with oxygen, one sees the rapid formation of a lethal cloud of nitrogen dioxide (NO_2) by the following reaction.

$$2\ NO + O_2 \rightarrow 2\ NO_2 \qquad \text{(Reaction 1)}$$

In a simple system, the nitrogen dioxide will combine reversibly with another nitric oxide to form dinitrogen trioxide, which adds water to form nitrite.

$$NO + NO_2 \leftrightarrow N_2O_3 \rightarrow 2\ NO_2^- + 2\ H^+$$

Nitrogen dioxide is a strong oxidant and certainly capable of oxidizing most biological molecules. But notice that the rate of reaction depends upon the square of nitric oxide concentration, so that the reactions slow down very rapidly as nitric oxide becomes dilute. Each 10-fold dilution slows the reaction a 100-fold. In the gas phase, pure nitric oxide is approximately 40 mM, and in saturated solutions of water, it is 1.9 mM. The physiological levels of nitric oxide range from approximately 10 nM, based upon the levels needed to cause vasodilation, to as high as 1-10 μM near fully activated macrophages.

If nitric oxide is only decomposing to form nitrogen dioxide, the half-life of 1 μM nitric oxide is approximately 11 minutes.[36] Because of the squared dependence on nitric oxide concentration, the half-life increases 10-fold for each 10-fold dilution. Consequently, the half-life of 10 nM nitric oxide is 1,100 minutes. These long half-lives of nitric oxide at dilute concentrations are why nitric oxide is far more likely to diffuse to a blood vessel and react with hemoglobin than to form nitrogen dioxide.

A subtle artifact of many in vitro experiments concerning the reactivity of nitric oxide is that no consideration is given to the reactions responsible for the removal of nitric oxide. Given many minutes of exposure, even 1 µM nitric oxide will react with oxygen to form nitrogen dioxide and thereby oxidize other molecules. Many in vitro studies add 100 µM to 1 mM concentrations of nitric oxide and conclude that it is toxic. However, diffusion through tissues to a blood vessel and reaction with hemoglobin will occur in seconds in vivo and thus safely remove nitric oxide produced at the levels used for signal transduction.

Many in vitro studies on the reactivity of nitric oxide are further complicated by the use of nitric oxide donors rather than working with nitric oxide itself. Paradoxically, the reactivity of many nitric oxide donors is greater than nitric oxide itself. For example, sodium nitroprusside generally oxidizes sulfhydryls to produce nitric oxide.[37] Nitrosothiols can rapidly modify protein thiols, whereas the equivalent reaction with nitric oxide under anaerobic conditions takes hours.[35] Thus, one must be cautious about studies that have used nitric oxide donors, because the actions may well not be due to nitric oxide itself.

A final admonition concerns the cross reactivity of cGMP with cAMP-dependent processes. Many in vitro experiments will utilize substantial amounts of a nitric oxide donor that will fully activate guanylate cyclase. Many cAMP-dependent processes can be activated by higher concentrations of cGMP, so that prolonged generation of cGMP may artifactually stimulate well known cAMP dependent processes.

REACTION WITH SUPEROXIDE

The production of nitric oxide by activated macrophages is important for their microbicidal and tumoricidal activities. One target is the inactivation of iron/sulfur centers. Because nitric oxide binds to metals, it is a logical conclusion that nitric oxide could directly inactivate these centers. However, more recent evidence suggests that nitric oxide itself is not responsible, but rather the formation of secondary oxidants is responsible.[38,39] Macrophages also reduce oxygen by one electron to produce superoxide as part of their microbicidal armamentarium. Superoxide reacts by a diffusion-limited reaction with nitric oxide to form a powerful and relatively stable oxidant, peroxynitrite.[40]

$$\cdot O_2^- + \cdot NO \rightarrow ONOO^-$$

The formation of peroxynitrite is nearly diffusion limited, occurring at a rate of 6.7×10^9 $M^{-1} \cdot s^{-1}$.[41] Superoxide dismutase, the principal scavenger of superoxide in vivo, reacts with superoxide at a rate of about 2×10^9 $M^{-1} \cdot s^{-1}$, which is the fastest rate of any enzyme but still 3-fold slower than superoxide reacting with nitric oxide. However, the intracellular concentration of superoxide dismutase is high, ranging from 2-40 μM or 0.1-0.5% of total soluble protein. Thus, the concentrations of nitric oxide normally produced for signal transduction (10-100 mM) will not effectively compete with superoxide dismutase for superoxide. On the other hand, pathological conditions can amplify the production of nitric oxide to the range of 1-10 μM. Consequently, nitric oxide is the only known biological molecule that can effectively compete with endogenous levels of superoxide dismutase for superoxide.[36]

Peroxynitrite is not a free radical because the unpaired electrons on superoxide and on nitric oxide have each combined to form a new chemical bond. At alkaline pH, peroxynitrite is an anion that is remarkably stable. Molar concentrations at pH 13 can be stored for days in the refrigerator and weeks in the freezer. We have recently shown that the unusual stability of peroxynitrite is due to its being bent into the *cis* conformation, which cannot be directly isomerized to the much more stable form, nitrate.[42] The pK_a of peroxynitrite is 6.8, so that at physiological pH approximately 20% of peroxynitrite will be protonated to form peroxynitrous acid. Peroxynitrous acid is a strong oxidant that can react with biological molecules by a number of complex mechanisms. It is particularly efficient at oxidizing iron/sulfur centers,[38,39] zinc fingers[43] and protein thiols.[44] It can also produce the products expected for hydroxyl radical attack. Thus, it can yield the hallmarks of oxidation used to implicate free radicals in vivo, including formation of protein carbonyls, salicylate hydroxylation and the formation of HO· products.[45,46] However, the direct reaction of peroxynitrite with specific moieties is more likely to account for its toxicity.[47] For example, incubation of *E. coli* with hydroxyl radical scavengers did not reduce the cytotoxicity of peroxynitrite.[48]

One important reaction of peroxynitrite is catalysis by transition metals, including the metal centers of superoxide dismutase and myeloperoxidase.[49] Transition metals catalyze a heterolytic cleavage to produce hydroxyl anion (HO⁻) plus nitronium ion (NO_2^+). Nitronium ion is well known to attack phenolics to produce nitrophenols. The nitration of protein tyrosine residues to give 3-nitrotyrosine is a footprint left by peroxynitrite in vivo. This reaction is catalyzed by superoxide dismutase, the enzyme that is also responsible for minimizing the formation of peroxynitrite by scavenging superoxide.[50]

We utilized the nitration of the low molecular weight phenolic 4-hydroxyphenylacetic acid catalyzed by superoxide dismutase as a trap to assay peroxynitrite formation by activated rat alveolar macrophages.[51] It is possible to eliminate 99% of the superoxide scavenging activity of superoxide dismutase by reacting it with alkaline hydrogen peroxide and deriving a critical arginine with phenylglyoxal. This modified form still catalyzes nitration with peroxynitrite. Freshly isolated rat alveolar macrophages were able to produce nitric oxide without stimulation but did not nitrate phenolics. Addition of phorbol esters stimulated superoxide production at a 4-fold greater rate than nitric oxide production. In the presence of native superoxide dismutase or the modified superoxide dismutase, the solution turned slightly yellow as the phenolic was nitrated. Quantitatively, all of the nitric oxide was converted to peroxynitrite, even with 500 U/ml of extracellular superoxide dismutase added. Thus, activated macrophages can produce substantial amounts of peroxynitrite when they are producing both superoxide and nitric oxide.

More recently, we have raised both polyclonal and monoclonal antibodies to nitrotyrosine-containing proteins. These can be used for both immunohistochemistry and for western blotting. We have shown that extensive nitration takes place around foamy macrophages in human atherosclerotic lesions.[52] Additional staining is found in foci of myocytes in the vascular smooth muscle. Staining of foamy macrophages was also observed in early fatty streaks.

A major concern with any antibody is its specificity. The nitrotyrosine antibody binding can be blocked by millimolar concentrations of nitrotyrosine or micromolar concentrations of short peptides containing nitrotyrosine. Thus, the antibody is better at recognizing nitrotyrosine in proteins than the free amino acid. Binding is not affected by tyrosine, chlorotyrosine, phosphotyrosine or aminotyrosine. Specificity can be further investigated by reducing the samples with dithionite, which will change nitrotyrosine to aminotyrosine. Thus, there are excellent controls that can be used even on human autopsy tissue to verify the presence of nitrotyrosine.

More recently, we have observed nitration in human lung biopsy and autopsy samples with sepsis, pneumonia or adult respiratory distress syndrome (ARDS).[53,54] Control lung samples show relatively little nitration. Exposure of rats to 100% oxygen will also induce nitration in lung alveoli at the time when pathology is apparent. Myocytes in human heart from patients with systemic sepsis or myocarditis show extensive nitration. By western blot analysis, actin appears to be nitrated in heart. It is possible that myocardial depression associated with nitric oxide production is due to nitration of the contractile protein machinery of muscle cells. Experimentally, actin has been artificially nitrated with tetranitromethane and shown to assemble abnormally.[55,56]

Tyrosine nitration is a convenient marker of reactive nitrogen-centered oxidants being produced. It is not necessarily due to the formation of peroxynitrite, but we have argued that peroxynitrite is the most probable source in vivo.[52] Other reactive species like nitrogen dioxide can form nitrotyrosine in simple solutions. Acidified nitrite can also produce nitrotyrosine if left for several days in contact with a protein. However, the amounts of nitrogen dioxide or nitrite that are present in vivo are far lower than necessary to cause significant nitration in vitro.

In a complex media such as the milieu of a cell, the reaction of superoxide dismutase with peroxynitrite becomes important because the superoxide dismutase combines rapidly with peroxynitrite and directs a more selective nitration of tyrosines on certain proteins. We have recently found that nitration of neurofilament L is greatly enhanced by superoxide dismutase. This may contribute to the role of superoxide dismutase in causing amyotrophic lateral sclerosis (ALS).[57] A series of 26 different dominant missense mutations at 18 different amin oacid positions to superoxide dismutase have been associated with the familial form of amyotrophic lateral sclerosis.[58] Expression of these mutations in transgenic mice causes development of motor neuron disease, whereas equivalent expression of the wild-type enzyme does not.[59] These results strongly suggest that an enhanced deleterious function of superoxide dismutase is responsible for the demise of motor neurons. Yet, superoxide dismutase is expressed in high concentrations in all cells, which makes the selective vulnerability of motor neurons difficult to explain.

We have proposed that one deleterious function of the mutant superoxide dismutases is the enhanced nitration of a key tyrosine kinase target.[57] However, nitration of neurofilaments may be a more likely explanation. The head domain of human neurofilament L contains ten tyrosines in the first ninety amino acids. Neurofilaments are intermediate filaments related to keratins that provide structural stability and control diameter in motor neurons.[60] Motor neurons are the largest neurons in the central nervous system and contain more neurofilaments than other cell types. Recently, the overexpression or mutation of neurofilaments has been shown to cause severe motor neuron injury in transgenic mice.[61,62] Nitration of tyrosines will change a normally hydrophobic residue into a negatively charged hydrophilic residue, and thereby disrupt the assembly of these proteins into a long polymeric structure. We have recently detected nitrated neurofilament L in autopsy tissue from ALS spinal cord patients. Nitration of cytoskeletal proteins is likely to occur in many other diseases because of their abundance and relative stability. Such modifications may affect cell morphology and adhesion.

In summary, the biological chemistry of nitric oxide can be grouped into three principal reactions: binding to ferrous heme in guanylate cyclase to activate cGMP, elimination by oxyhemoglobin and formation of toxic oxidants by reacting with superoxide. These three reactions are certainly a simplification but a reasonable summary because they are rapid under physiologically relevant conditions. The diffusion of nitric oxide has been often overlooked in understanding the biological functioning of nitric oxide as a messenger.[12] More detailed reviews of nitric oxide chemistry and synthesis have been reported.[24,63] We have not considered the chemistry of nitrosylated derivatives, though nitrosothiols are found in submicromolar amounts in normal plasma and tissues with levels rising under pathological conditions.[64]

ACKNOWLEDGMENTS

This work was supported by grants HL46407, NS24338, and HL48676 from the National Institutes of Health and from the American Heart Association. J.S. Beckman is supported as an Established Investigator of the American Heart Association.

REFERENCES

1. Moncada S, Palmer RMJ, Higgs EA. Nitric oxide: physiology, pathophysiology, and pharmacology. Pharmacol Rev 1991; 43:109-142.
2. Iadecola C, Pelligrino DA, Moscowitz MA, Lassen NA. State of the Art Review: Nitric oxide inhibition and cerebrovascular regulation. J Cereb Blood Flow Metabol 1994; 14:175-192.
3. Jones K. The chemistry of nitrogen. Oxford, Pergamon Press, 1973.
4. Traylor TG, Sharma VS. Why NO? Biochemistry 1992; 31:2847-2849.
5. Ignarro LJ. Heme-dependent activation of soluble guanylate cyclase by nitric oxide: regulation of enzyme activity by porphyrins and metalloporphyrins. Sem Hematology 1989; 26:63-76.
6. Doyle MP, Hoekstra JW. Oxidation of nitrogen oxides by bound dioxygen in hemoproteins. J Inorg Biochem 1981; 14:351-358.
7. Green LC, Wagner DA, Glogowski J, Skipper PL, Wishnok JS, Tannenbaum SR. Analysis of nitrate, nitrite, and [15N]nitrate in biological fluids. Anal Biochem 1982; 126:131-138.
8. Wise DL, Houghton G. Diffusion of nitric oxide. Chem Eng Sci 1968; 23:1211-1216.
9. Meulemans A. Diffusion coefficients and half-lifes of nitric oxide and N-nitrosoarginine in rat cortex. Neurosci Lett 1994; 171:89-93.
10. Kelm M, Schrader J. Nitric oxide release from the isolated guinea pig heart. Eur J Pharmac 1988; 155:317-321.
11. Beckman JS, Chen J, Crow JP, Ye YZ, ed. Reactions of nitric oxide, superoxide and peroxynitrite with superoxide dismutase in neurodegeneration. Progress in Neurology, 1994.
12. Lancaster JR. Simulation of the diffusion and reaction of endogenously produced nitric oxide. Proc Natl Acad Sci USA 1994; 91:8137-8141.
13. Griffith TM, Edwards DH, Davies RL, Harrison TJ, Evans KT. EDRF coordinates the behaviour of vascular resistance vessels. Nature, London 1987; 329:442-445.
14. Schuman EM, Madison DV. Locally distributed synaptic potentation in the hippocampus. Science 1994; 263:532-536.
15. Gaily JA, Montague PR, Reeke Jr GN, Edelman GM. The NO hypothesis: Possible effects of a short-lived, rapidly diffusible signal in the development and function of the nervous system. Proc Natl Acad Sci (USA) 1990; 87:3547-3551.
16. Montague PR, Gally JA, Edelman GM. Spatial signaling in the development and function of neural connections. Cereb Cortex 1991; 1:199-220.
17. Deguchi T, Yoshioka M. L-Arginine identified as an endogenous activator for soluble guanylate cyclase from neuroblastoma cells. J Biol Chem 1982; 257:10147-10151.
18. Granger DL, Hibbs Jr JB, Perfect JR, Durack DT. Specific amino acid (L-arginine) requirement for the microbiostatic activity of mu-

rine macrophages. J Clin Invest 1988; 81:1129-1136.

19. Bredt DS, Synder HS. Isolation of nitric oxide synthetase, a calmodulinrequiring enzyme. Proc Natl Acad Sci (USA) 1990; 87:682-685.

20. Bredt DS, Hwang PM, Glatt CE, Lowenstein C, Reed RR, Snyder SH. Cloned and expressed nitric oxide synthase structurally resembles cytochrome P450 reductase. Nature 1991; 351:714-718.

21. Stuehr DJ, Kwon NS, Nathan CF, Griffin OW, Feldman PL, Wiseman JN. Why droxy-L-arginine is an intermediate in the biosynthesis of nitric oxide from L-arginine. J Biol Chem 1991; 266:6259-6263.

22. Stuehr DJ, Ikeda-Saito M. Spectral characterization of brain and macrophage nitric oxide synthases. J Biol Chem 1992; 267: 20547-20550.

23. Hevel JM, Marletta MA. Macrophage nitric oxide synthase: Relationship between enzyme-bound tetrahydrobiopterin and synthase activity. Biochemistry 1992;31:7160-7165.

24. Feldman PL, Griffith OW, Stuehr DJ. The surprising life of nitric oxide. Chem Eng News 1993; 71:26-38.

25. Stuehr DJ, Nathan CF. Nitric oxide. A macrophage product responsible for cytostasis and respiratory inhibition in tumor cells. J Exp Med 1989; 169:1543-1555.

26. Radomski MW, Palmer RMJ, Moncada S. Glucocorticoids inhibit the expression of an inducible, but not the constitutive, nitric oxide synthase in vascular endothelial cells. Proc Natl Acad Sci USA 1990; 87:10043-10047.

27. Koprowski H. Zheng YM, Heber-Katz E, Fraser N, Rorke L, Fu ZF, Hanlon C, Dietzschold B. In vivo expression of inducible nitric oxide synthase in experimentally induced neurological diseases. Proc Natl Acad Sci USA 1993; 90:3024-3027.

28. Hibbs Jr JB, Westenfelder C, Taintor R, Vavrin Z, Kablitx C, Baranowski RL, Ward JH, Menlove RL, McMurry MP, Kushner JP, Samlowski WE. Evidence for cytokine-inducible nitric oxide synthesis from L-arginine in patients receiving interleukin-2 therapy. J Clin Invest 1992; 89:867-877.

29. Nathan C, Xie QW. Regulation of biosynthesis of nitric oxide. J Biol Chem 1994; 269:13725-13278.

30. Karupiah G, Xie Q, Buller ML, Nathan C, Duarte C, MacMicking JD. Inhibition of viral replication by interferon-gamma-induced nitric oxide synthase. Science 1993; 261:1445-1448.

31. Kwon NS, Stuehr DH, Nathan CF. Inhibition of tumor cell ribonucleotide reductase by macrophage-derived nitric oxide. J Exp Med 1991; 174:761767.

32. Lepoivre M, Flaman J-M, Henry Y. Early loss of the tyrosyl radical in ribonucleotide reductase of adenocarcinoma cells producing nitric oxide. J Biol Chem 1992; 267:22994-23000.

33. Albina JE, Cui S, Mateo RB, Reichner S. Nitric oxide-mediated apoptosis in murine peritoneal macrophages. J Immunol 1993; 150:5080-5085.
34. Cui S, Reichner JS, Mateo R, Albina J. Activated murine machophages induce apoptosis in tumor cells through nitric oxide-dependent or independent mechanisms. Cancer Res 1994; 54: 2462-2467.
35. Pryor WA, Church DF, Govindan CK, Crank G. Oxidation of thiols by nitric oxide and nitrogen dioxide: synthetic utility and toxicological implications. J Org Chem 1982; 147:156-158.
36. Beckman JS, Tsai JHM. Reaction rates and diffusion in the toxicity of peroxynitrite. The Biochemist 1994; 16:8-10.
37. Bates JN, Baker MT, Guerra Jr R, Harrison DG. Nitric oxide generation from nitroprusside by vascular tissue. Evidence that reduction of the nitroprusside anion and cyanide loss are required. Biochem Pharmacol 1991; 42:S157S165.
38. Castro L, Rodriguez M, Radi R. Aconitase is readily inactivated by peroxynitrite, but not by its precurser, nitric oxide. J Biol Chem 1994; 269:29409-29415.
39. Hausladen A, Fridovich I. Superoxide and peroxynitrite inactivate aconitases, nitric oxide does not. J Biol Chem 1994; 269:2940529408.
40. Blough NV, Zafiriou OC. Reaction of superoxide with nitric oxide to form peroxonitrite in alkaline aqueous solution. Inorg Chem 1985; 24:35043505.
41. Huie RE, Padmaja S. The reaction rate of nitric oxide with superoxide. Free Rad Res Commun 1993; 18:195-199.
42. Tsai JHM, Hamilton TP, Harrison JG, Jablowski M, Woerd MVD, Martin JC, Beckman JS. Role of peroxynitrite conformation with its stability and toxicity. J Amer Chem Soc 1994; 116:4115-4116.
43. Crow JP, McCord JM, Beckman JS. Oxidative inactivation of the zinc-thiolate in yeast. Biochemistry 1995; In press.
44. Radi R, Rodriguez M, Castro L, Telleri R. Inhibition of mitochondrial electron transport by peroxynitrite. Arch Biochem Biophys 1994; 308:8995.
45. Beckman JS, Beckman TW, Chen J, Marshall PM, Freeman BA. Apparent hydroxyl radical production from peroxynitrite: implications for endothelial injury by nitric oxide and superoxide. Proc Natl Acad Sci (USA) 1990; 87:1620-1624.
46. Crow JP, Spruell C, Chen J, Gunn C, Ischiropoulos H, Tsai M, Smith CD, Radi R, Koppenol WH, Beckman JS. On the pH-dependent yield of hydroxyl radical products from peroxynitrite. Free Radical Biol Med 1994; 16:331338.
47. Koppenol WH, Moreno JJ, Pryor WA, Ischiropoulos H, Beckman JS. Peroxynitrite: a cloaked oxidant from superoxide and nitric oxide. Chem Res Toxicol 1992; 5:834-842.

48. Zhu L, Gunn C, Beckman JS. Bactericidal activity of peroxynitrite. Arch Biochem Biophys 1992; 298:452-457.

49. Ischiropoulos H, Zhu L, Chen J, Tsai HM, Martin JC, Smith CD, Beckman JS. Peroxynitrite-mediated tyrosine nitration catalyzed by superoxide dismutase. Arch Biochem Biophys 1992; 298:431-437.

50. Beckman JS, Ischiropoulos H, Zhu L, van der Woerd M, Smith C, Chen J, Harrison J, Martin JC, Tsai M. Kinetics of superoxide dismutase and iron catalyzed nitration of phenolics by peroxynitrite. Arch Biochem Biophys 1992; 298:438445.

51. Ischiropoulos H, Zhu L, Beckman JS. Peroxynitrite formation from activated rat alveolar macrophages. Arch Biochem Biophys 1992; 298:446-451.

52. Beckman JS, Ye YZ, Anderson P, Chen J, Accavetti MA, Tarpey MM, White CR. Extensive nitration of protein tyrosines observed in human atherosclerosis detected by immunohistochemistry. Biological Chemistry Hoppe-Seyler 1994; 375:81-88.

53. Haddad I, Pataki G, Hu P, Galliani C, Beckman JS, Matalon S. Quantitation of nitrotyrosine levels in lung sections of patients and animals with acute lung injury. J Clin Inv 1994.

54. Kooy NW, Royall JA, Ye YZ, Kelley DR, Beckman JS. Evidence for in vivo peroxynitrite production in human acute lung injury. Amer Rev Respir Dis 1994; Submitted.

55. Chantler PD, Gratzer WB. Effects of specific chemical modification of actin. Eur J Biochem 1975; 60:67-72.

56. Miki M, Barden JA, dos Remedios CG, Phillips L, Hambly BD. Interaction of phalloidin with chemically modified actin. Eur J Biochem 1987; 165:125130.

57. Beckman JS, Carson M, Smith CD, Koppenol WH. ALS, SOD and Peroxynitrite. Nature 1993; 364:584.

58. Rosen DR, Siddique T, Patterson D, Figlewicz DA, Sapp P, Hentati A, Donaldson D, Goto J, O'Regan JP, Deng H-X, Rahmani Z, Krizus A, McKenna-Yasek D, Cayabyab A, Gaston SM, Berger R, Tanszi RE, Halperin JJ, Herzfeldt B, Van den Bergh R, Hung W-Y, Bird T, Deng G, Mulder DW, Smyth C, Lang NG, Soriana E, Pericak-Vance MA, Haines J, Rouleau GA, Gusella JS, Horvitz HR, Brown RHJ. Mutations in Cu/Zn superoxide dismutase gene are associated with familial amyotrophic lateral sclerosis. Nature 1993; 362:59-62.

59. Gurney ME, Pu H, Chiu AY, Dal Corto MC, Polchow CY, Alexander DD, Caliendo J, Hentati A, Kwon YW, Deng H-X, Chen W, Zhai P, Sufit RL, Siddique T. Motor neuron degeneration in mice that express a human Cu,Zn superoxide dismutase mutation. Science 1994; 264:1772-1774.

60. Brady ST. Motor neurons and neurofilaments in sickness and in health. Cell 1993; 73:1-3.

61. Xu Z, Cork L, Griffin J, Cleveland D. Increased expression of neurofilament subunit NF-L produces morphological alterations that

resemble the pathology of human motor neuron disease. Cell 1993; 73:23-33.

62. Lee MK, Marszalek JR, Cleveland DW. A mutant neurofilament subunit causes massive, selective motor neuron: Implications for the pathogenesis of human motor neuron disease. Neuron 1994; 13:975-988.

63. Stamler JS, Singel DJ, Loscalzo J. Biochemistry of nitric oxide and its redox activated forms. Science 1992; 258:1898-1902.

64. Stamler JS, Simon DI, Osborne JA, Mullins ME, Jaraki O, Michel T, Singel DJ, Loscalzo J. S-Nitrosylation of proteins with nitric oxide: Synthesis and charactization of biologically active compounds. Proc Natl Acad Sci USA 1992; 89:444-448.

CHAPTER 2

NITRIC OXIDE: A HOMEOSTATIC REGULATOR OF LEUKOCYTE-ENDOTHELIAL CELL INTERACTIONS

Paul Kubes

INTRODUCTION

Inflammation is a localized response induced by microbial infection and/or cellular and tissue injury. Inadvertent activation of the inflammatory cascade via complement activation, enhanced arachidonate metabolism and cytokine production may result in 1) abnormal recognition of the host as "foreign" or 2) absence of a "termination" signal in an otherwise normal inflammatory process. This is thought to lead to acute inflammation including reperfusion injury as well as chronic inflammatory processes such as debilitating joint injury, inflammatory bowel disease and complications associated with sepsis. Central to the onset of these disease states is the accumulation of leukocytes which cause vascular and tissue injury by virtue of their ability to synthesize and release large quantities of reactive oxygen metabolites and proteases. For leukocytes to gain access to a potential site, they must leave the mainstream of blood, make initial contact or tether to the endothelium and then roll along the length of postcapillary venules before coming to a firm adhesion and final emigration out of the vasculature. Although much emphasis has been placed on the identification and characterization of cellular and molecular mechanisms that promote leukocyte recruitment to target sites of inflammation, very little is known about the potential role of endogenous anti-inflammatory and antiadhesion molecules.

Nitric oxide (NO) is a biologically active molecule continuously produced by the lining of vessels (endothelium), that in addition to maintaining vascular perfusion in vivo, has profound anti-aggregatory and antiadhesive properties for leukocytes and platelets, respectively, in vitro. Based on these simple observations, approximately 5 years ago, the hypothesis was put forth that NO may be an endogenous homeostatic regulator of leukocyte-endothelial cell interactions in the microcirculation. In this review I will summarize the evidence to support the contention that NO is an endogenous antiadhesive molecule for leukocytes and highlight some of the areas of controversy and uncertainty regarding this thesis.

MECHANISMS OF LEUKOCYTE RECRUITMENT

Figure 2.1 briefly summarizes the multistep cascade of events that leads to leukocyte infiltration to sites of inflammation and injury. A leukocyte moving at very high speeds in the mainstream of blood first makes contact with the endothelial cells lining the vessel wall (tethering) and then moves along the endothelium at a greatly reduced velocity relative to red blood cells (rolling). These two events are thought to be dependent primarily upon three members of the selectin family of adhesion molecules termed L-selectin, P-selectin and E-selectin. Whereas L-selectin is constitutively expressed on leukocytes, the expression of P-selectin (induced in minutes) and E-selectin (4-6 hrs for maximal induction) is regulated on the endothelium.

P-selectin, is a very likely candidate as a mediator of the early phase of leukocyte rolling. This selectin is stored in Weibel-Palade bodies of endothelial cells and is rapidly expressed on the cell surface by histamine and other mediators.[1] Histamine-induced, P-selectin expression supports transient leukocyte adhesion to endothelial cells in static assay systems.[1] Moreover, Lawrence and Springer[2] demonstrated that leukocytes exposed to shear forces rolled on artificial lipid bilayers containing purified P-selectin but not other adhesion molecules (ICAM-1). A recent publication has

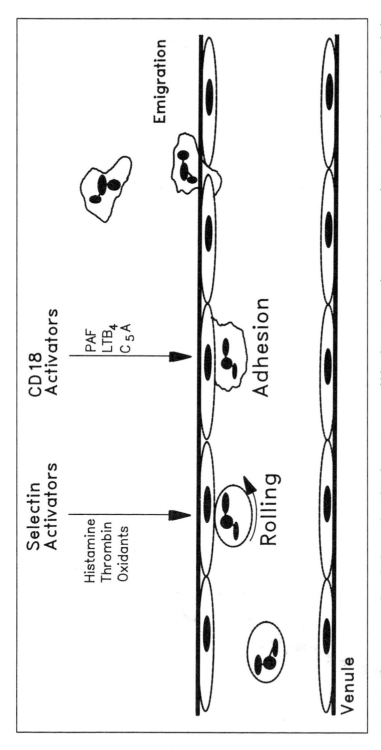

Fig. 2.1. Leukocytes move from high speeds within the mainstream of blood to extravascular space via a multistep series of events that include selectin-dependent rolling and β₂-integrin (CD18)-dependent adhesion and emigration. Histamine, thrombin and oxidants can activate P-selectin-dependent rolling whereas cytokines (not shown) can activate E-selectin-dependent rolling. Chemotactic agents (PAF, LTB₄, C₅A) activate CD18-dependent adhesion and emigration.

shown that exposure of human umbilical vein endothelial cells to histamine causes P-selectin-dependent leukocyte rolling under shear conditions in vitro.[3] These in vitro studies would support the hypothesis that histamine can induce P-selectin on endothelium and thereby promote leukocyte rolling. This past year, two studies have described the fact that histamine induces leukocyte rolling in vivo, an event that could be entirely prevented by an anti-P-selectin antibody.[4,5]

Much of the evidence that implicates E-selectin as a mediator of leukocyte rolling in inflamed microvessels is inferred from in vitro studies. E-selectin is only expressed after a few hours on the surface of endothelium exposed to LPS, IL-1, TNFα and perhaps other cytokines suggesting that this selectin contributes to the later recruitment of leukocytes.[6] E-selectin expressed on transfected L-cells, endothelial cell monolayers treated with IL-1, TNFα or LPS, and substrates bearing purified E-selectin all supported neutrophil rolling in flow chambers.[7] The dearth of information regarding the contribution of E-selectin to leukocyte-endothelial cell interactions in vivo results from the short-term experimental protocols that have dominated published work employing intravital videomicroscopy. Olofsson et al[8] have recently reported that IL-1 treatment of rabbit mesenteries supported leukocyte rolling via E-selectin at 4 hrs of cytokine administration. However, E-selectin-dependent leukocyte rolling in postcapillary venules in various inflammatory models has not been examined to date.

Finally, L-selectin has also been demonstrated to mediate leukocyte rolling in vitro and in vivo. A monoclonal antibody directed against L-selectin (DREG-200) dramatically reduced the number of rolling leukocytes in rabbit mesenteric venules whereas soluble recombinant immunoglobulin G chimera or polyclonal antiserum to L-selectin inhibited leukocyte rolling in rat mesenteric tissue.[9,10] Flow chamber studies have demonstrated that L-selectin antibodies will significantly prevent leukocyte rolling on endothelium. Recently, Lawrence et al[11] reported that when L-selectin is shed from leukocytes these cells if allowed to settle on E-selectin-coated surfaces would subsequently roll. However, the initial ability to tether to E-selectin, that appears to be essential for the subsequent rolling, was lost. These data suggest that L-selectin on leukocytes may be essential for the initial contact, whereas the endothelial selectins support the rolling.

Once a cell begins to roll, it can then firmly adhere and finally emigrate out of the vasculature. It should be noted that this is an interdependent series of events inasmuch as inhibiting leukocyte rolling prevents subsequent leukocyte adhesion and ultimately leukocyte emigration out of the vasculature.[9,10] The firm adhesion is mediated by the integrins found on leukocytes, and in the case of the neutrophil, the β_2-integrin (CD11/CD18). The significance of neutrophil CD18 function is exemplified by the observations with neutrophils from patients congenitally lacking the β_2 integrins; these cells are unable to adhere to biologic substratum and do not accumulate at inflammatory sites.[12] Leukocyte recruitment to acute inflammatory sites has also been almost entirely impaired by monoclonal antibodies against CD18, and the inability of leukocytes to adhere rather than roll was the underlying mechanism.[13] Although CD18 can bind to multiple ligands, the endothelial-derived intracellular adhesion molecules (ICAM-1 and ICAM-2) form important adhesive interactions with CD18. Some excellent reviews on the underlying adhesive mechanisms for leukocyte-endothelial cell interactions have been published.[14-16]

Much of the work in vivo to study leukocyte rolling and adhesion is based on intravital videomicroscopy which allows investigators to visualize leukocyte-endothelial cell interactions within the microcirculation (Fig. 2.2). Under normal conditions, in 20-40 μm postcapillary venules only a few leukocytes can be seen rolling and adhering to the vessel wall (Fig. 2.2A). Following the induction of an inflammatory response the number of rolling and adhering cells is greatly increased (Fig. 2.2B). This technique has been the basis for the work summarized in this chapter and some of the other chapters in this book.

NITRIC OXIDE INHIBITION
CAUSES LEUKOCYTE ROLLING AND ADHESION

Nitric oxide continuously produced from L-arginine via the constitutively expressed enzyme, NO synthase, was first discovered as an endothelium-derived relaxing factor of the vasculature.[17] However, since the advent of NO synthesis inhibitors including N^G-monomethyl L-arginine (L-NMMA) and L-N^G-nitro-arginine-methyl-ester (L-NAME), it became clear that endogenous NO regulated numerous biologic responses within the microcirculation. Initial in vitro work demonstrated that NO prevented neutrophil

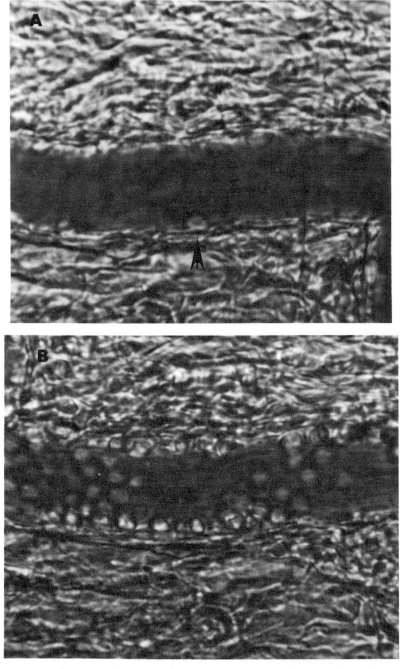

Fig. 2.2. Micrographs illustrating a single 35 μm venule under normal conditions and following an inflammatory insult as seen online using intravital microscopy. In this experiment the vessel was exposed to 1 hr of ischemia and 3 mins of reperfusion. The arrow indicates a rolling leukocyte under control conditions (Panel A) versus many rolling and adhering leukocytes (Panel B). See chapter 5 for description of ischemia/reperfusion. Similar images are obtained when the tissue is superfused with an NO-synthesis inhibitor.

aggregation,[18] suggesting that the cell-cell interactions underlying leukocyte adhesion might be regulated by nitric oxide. Based on this initial observation, we asked the question, can inhibition of endogenous NO synthesis modulate leukocyte function in vivo? Using the intravital microscopy preparation described above, the cat mesenteric microcirculation was superfused with L-NAME and observed for 60 mins.[19] Although blood pressure rose rapidly and intestinal blood flow began to decrease almost immediately, there was no evidence of increased leukocyte adhesion within the microcirculation for the first 15-20 mins. However, at 30 mins a very profound increase in leukocyte adhesion was observed in postcapillary venules but not in precapillary arterioles. This time course is summarized in Figure 2.3.

One potential mechanism to explain the enhanced leukocyte influx with L-NAME on the venular side could be related to the reduced blood flow and associated hemodynamic forces that normally push leukocytes through blood vessels. As hemodynamic forces are reduced with the addition of L-NAME, leukocyte adhesion is favored. Moreover, since the shear forces are so high on the arteriolar side the L-NAME-induced reduction in shear forces may have been insufficient to cause leukocyte adhesion on the precapillary side of the microvasculature. We examined the adhesion responses elicited by a step-wise reduction in blood flow in the presence and absence of nitric oxide (L-NAME superfusion) to determine whether a reduction in hydrodynamic dispersal forces could account for the high adhesion with L-NAME. The data (Fig. 2.4) revealed that the leukocyte adhesion on the venular side was significantly higher at a given shear rate with L-NAME than without the NO synthesis inhibitor.[19] This observation indicates that shear rate-dependent leukocyte adhesion cannot account for the majority of adhering leukocytes on the venular side associated with inhibition of NO production. Also noteworthy is the fact that reduction in shear rates within arterioles to levels equivalent to or lower than those in venules did not induce any notable leukocyte-endothelial cell interactions with or without L-NAME.

To establish that the L-NAME-induced leukocyte adhesion was due to inhibition of NO and not nonspecific effects, L-NAME-treated animals were given large amounts of L-arginine or the biologically inactive enantiomer, D-arginine.[19] L-arginine not D-arginine reversed the L-NAME-induced adhesion, suggesting that this

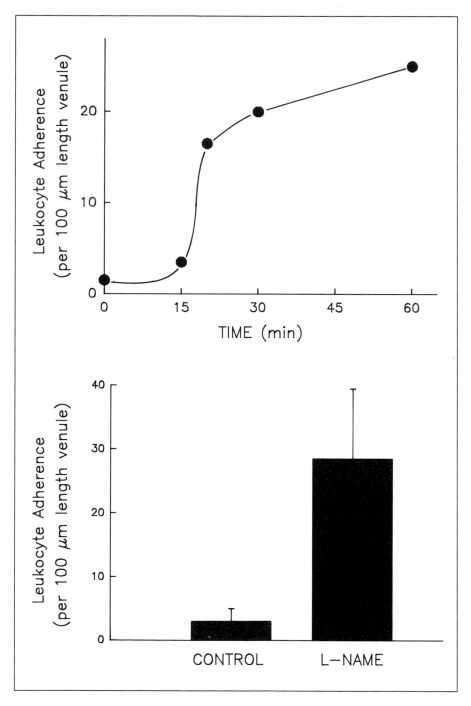

Fig. 2.3. A time course of leukocyte adhesion in postcapillary venules exposed to a nitric oxide synthase inhibitor for the first 60 mins (top panel). A rapid onset of leukocyte adhesion was observed at 20-30 mins of L-NAME. The lower panel illustrates cumulative data illustrating a 10- to 15-fold increase in leukocyte adhesion. A very significant proportion of the adherent leukocytes emigrate out of the vasculature at 60 mins of L-NAME exposure.

was indeed an effect of NO synthesis inhibition. Moreover, addition of exogenous sources of NO (nitroprusside) or supplementing tissue with cGMP analogs prevented the leukocyte adhesion,[20,21] raising the possibility that NO and the intracellular messenger, cGMP, are homeostatic regulators of leukocyte adhesion. A second NO synthesis inhibitor, L-NMMA,[19] — as well as an endogenous NO inhibitor, dimethyl-L-arginine,[22] thought to play a role in atherosclerosis (see chapter 6) — also caused a rapid increase in leukocyte adhesion within 30 mins. Since that initial observation that NO synthesis inhibition caused leukocyte adhesion in the cat mesentery, similar data have been reported in rat mesenteric venules,[20,23,24] rat lung vasculature,[25] the cat heart vasculature,[26] rat skeletal muscle postcapillary venules[27] and mouse liver sinusoids.[28] Finally, aminoguanidine, the selective inhibitor of iNOS, has also been shown to cause leukocyte adhesion, however whether this is due to its overlapping ability to inhibit constitutive nitric oxide synthase remains unknown.[29]

Leukocyte adhesion induced by various proinflammatory molecules including platelet-activating factor and LTB$_4$ is primarily mediated via the leukocyte CD11/CD18 glycoprotein complex.[30] Experiments on the cat mesentery revealed that this mechanism was also responsible for L-NAME-induced leukocyte adhesion.[19] A monoclonal antibody directed against the CD11/CD18 glycoprotein complex (MoAB IB$_4$) that immunoneutralizes this adhesion molecule prevented essentially all of the L-NAME-induced adhesion. However, unlike PAF or LTB$_4$, L-NAME did not directly increase the expression or adhesiveness of CD11/CD18 in vitro.[19,23] Clearly, L-NAME per se did not activate CD11/CD18 on leukocytes. At the time the lack of effect of L-NAME on neutrophil adhesion in vitro was postulated to be due to the absence of some factor and/or cell that modulates the L-NAME-induced leukocyte adhesion in vivo.

In addition to leukocyte adhesion, endogenous NO production has also been shown to play a role in the regulation of P-selectin-dependent leukocyte rolling in the rat mesenteric microvasculature. Pretreatment of animals with L-NAME caused increased leukocyte rolling in a rapid manner consistent with P-selectin expression.[31] Moreover, immunohistochemistry revealed that L-NAME caused a significant increase in P-selectin expression in postcapillary venules of the rat microvasculature. The

increased leukocyte rolling in venules without NO production could
be entirely inhibited by a monoclonal antibody directed against
P-selectin suggesting that the increased expression of this adhesion
molecule mediates the increase in leukocyte rolling associated with
NO synthesis inhibition. The increased level of P-selectin expression and leukocyte rolling could be greatly reduced if a cGMP
analog was added prior to superfusion of the rat mesentery with
L-NAME. These data suggest that cGMP may regulate P-selectin
expression on endothelium, a thesis consistent with the ability of
elevated levels of cGMP to inhibit surface expression of P-selectin
on platelets.[32] Whether L-NAME will promote P-selectin expression and leukocyte rolling on endothelium in vitro in flow chambers remains unknown to date. As previously described, we have
demonstrated that L-NAME will not cause adhesion in vitro, however static assay systems not flow chambers were used, and con-
clusions about leukocyte rolling from this type of experiment can-

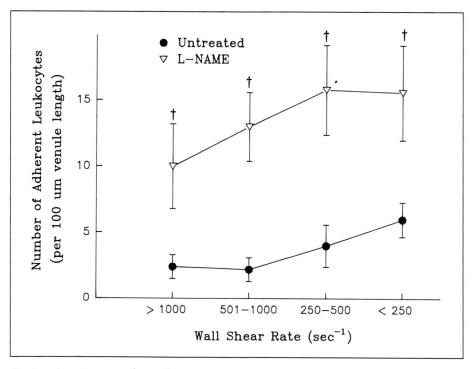

Fig. 2.4. One important factor that can regulate leukocyte adhesion to a vessel is the speed
of red blood cells moving through that vessel. The shear forces (proportional to red blood cell
velocity) were decreased following L-NAME infusion. However, when blood flow was reduced
mechanically in the presence or absence of L-NAME, at every shear rate, L-NAME administration
caused a significantly higher degree of leukocyte adhesion than in untreated animals.
† p < 0.05 relative to respective untreated value.

not be made. Therefore, whether L-NAME was acting directly on endothelium to inhibit NO or on some other cell remains unknown.

MAST CELLS
AND NITRIC OXIDE SYNTHESIS INHIBITION

There is a growing body of evidence to suggest that exogenous NO could diminish mast cell reactivity. Salvemini et al[33] reported that nitroprusside decreased the amount of histamine released by mast cells. Moreover, this group demonstrated that L-NAME augmented the release of histamine from mast cells. Hogaboam et al[34] reported that nitric oxide inhibited platelet-activating factor (PAF) production from mast cells. The relevance of these data to leukocyte recruitment becomes obvious when one considers the close proximity of mast cells to the vasculature in various tissues and the fact that mast cells contain a myriad of proinflammatory molecules capable of recruiting leukocytes.[35] Activated mast cells release preformed molecules within granules (chemotactic peptides, proteases and histamine), newly synthesized mediators (leukotrienes, PAF and prostaglandins) and various cytokines (TNFα and IL-1). It is therefore not surprising that activation of mast cells causes leukocyte recruitment both at an early time point (30-60 mins) via rapid induction of P-selectin[36] and in more delayed-type late phase reactions (4-8 hrs) via induction of E-selectin.[37] Therefore, it seems reasonable that if nitric oxide prevents the release of mediators from mast cells, that this process may play an important role in preventing mast cell-dependent leukocyte recruitment.

Very recent, unpublished data from our laboratory would suggest that nitric oxide donors may attenuate mast cell-dependent leukocyte recruitment in vivo and in vitro (Fig. 2.5). Superfusion of the pharmacological reagent, compound 48/80 onto the rat mesentery, caused mast cell degranulation and leukocyte rolling via a histamine and P-selectin-dependent event. Moreover, a significant increase in leukocyte adhesion was observed that was dependent on PAF and CD18.[38] The increased rolling and adhesion was almost entirely prevented by the nitric oxide donor spermine-NO. Moreover, in vitro work has revealed that in a static assay system, compound 48/80 caused mast cell-dependent leukocyte adhesion that was inhibited by the PAF receptor antagonist (WEB 2086). Spermine-NO in this assay system also prevented the mast

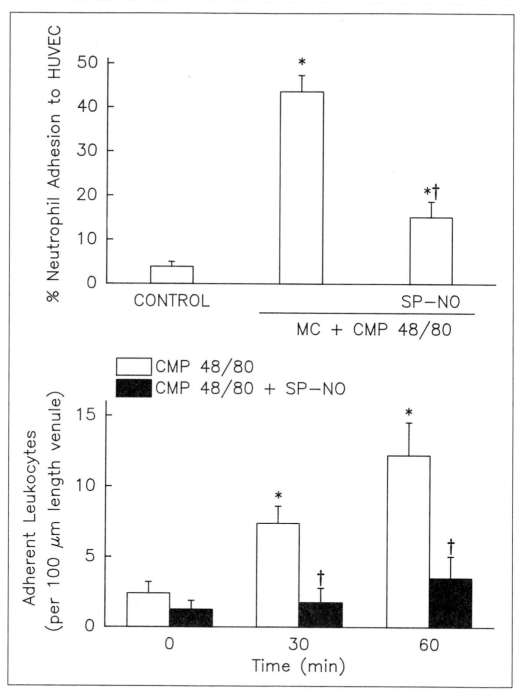

Fig. 2.5. The top panel illustrates that the mast cell degranulator compound 48/80 evokes leukocyte adhesion in vitro to endothelium in the presence (not absence; not shown) of mast cells. This effect can be entirely inhibited by either an NO donor (spermine-NO) or a PAF-receptor antagonist (WEB 2086; not shown). Direct application of PAF caused increased adhesion that was not inhibitable with spermine-NO (not shown). The bottom panel demonstrated that compound 48/80-induced leukocyte adhesion in vivo was inhibited by spermine-NO (unpublished observations). *p < 0.05 relative to control or time 0. † p < 0.05 relative to CMP 48/80 value.

cell-dependent (PAF-dependent) adhesion, but spermine-NO was unable to inhibit leukocyte adhesion directly by PAF. The only explanation available is that spermine-NO interferes with the release of, or synthesis of, PAF from the mast cell. This is supported by the fact that once the endothelium is exposed to PAF, spermine-NO has little effect on the process. These data raise the possibility that nitric oxide donors may have potential therapeutic properties in mast cell-mediated diseases such as anaphylaxis, asthma and perhaps even arthritis.

Removal of endogenous NO seemed to activate mast cells in vivo (Fig. 2.6). Studies from our laboratory demonstrated that connective tissue mast cells located in close proximity to the mesenteric vasculature and capable of releasing a myriad of proinflammatory molecules[35,39] became activated following nitric oxide synthesis inhibition.[23] In another study mast cell-derived protease II levels from mucosal mast cells were shown to be elevated following systemic administration of L-NAME,[40] suggesting that both connective tissue mast cells (from mesentery) and mucosal mast cells (intestinal mucosa) may be under physiologic control from the continuous release of endogenous nitric oxide. These results were extended by two groups; exposure of tissue to L-NAME caused detectable increase in oxidative stress within mast cells in vivo.[24,41] Moreover, various mast cell stabilizers prevented the L-NAME-induced leukocyte recruitment suggesting that mast cells played an important role in leukocyte adhesion following nitric oxide synthesis inhibition.[23] In addition, a PAF-receptor antagonist (WEB 2086) and an LTB$_4$-receptor antagonist (SC 41930) reduced the increased leukocyte adhesion associated with L-NAME.[42] These data suggest that the increased mast cell-dependent leukocyte adherence involved PAF and LTB$_4$. Although the source of these chemoattractants remains unknown, it is tempting to speculate that inhibition of nitric oxide synthesis activates mast cells to release PAF and LTB$_4$ and induce leukocyte adhesion. It is noteworthy that mast cells can synthesize both PAF[43] and LTB$_4$[39] and that in for example intestinal tissues, mast cells are the primary source of PAF.[43]

NITRIC OXIDE SYNTHESIS INHIBITION AND THE ROLE OF OXIDANTS

Based on previous work that superoxide could avidly inactivate endothelium-derived NO,[44] the antithesis that NO may be

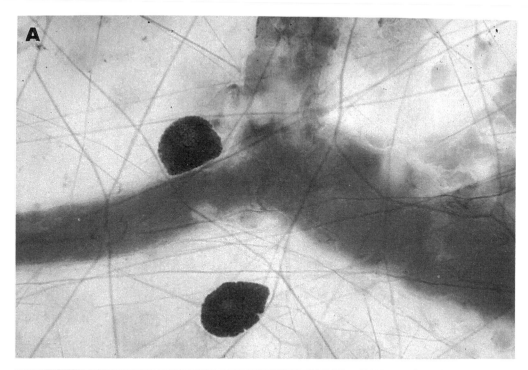

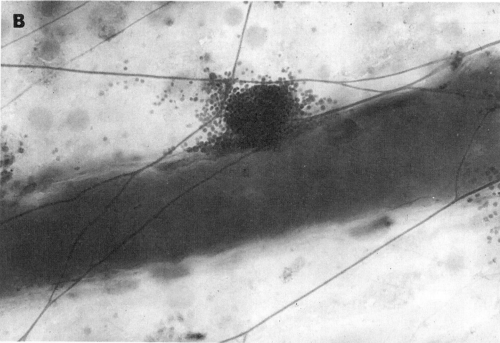

Fig. 2.6. Panel A illustrates two intact mast cells (red granule cells) in close proximity to a rat mesenteric vessel (orange due to safranine staining). Panel B demonstrates a mast cell degranulating after exposure to L-NAME. Some cells remain intact whereas others partly or fully degranulate following L-NAME treatment. (Note: original color photos were converted to black and white greyscale. Orange will appear as medium grey and red will appear as dark grey.)

an endogenous scavenger of superoxide was proposed by numerous laboratories. This idea was appealing inasmuch as superoxide has been shown to directly cause leukocyte rolling and adhesion, and superoxide is responsible for leukocyte recruitment in ischemia/reperfusion and various other disease states.[45-48] Superoxide is generated continuously by virtually all cells; 1-4% of oxygen used in the mitochondrial electron transport pathway results in the production of superoxide.[49] If endogenous NO contributes to the removal of endogenously produced superoxide, then diminishing the former could conceivably lead to elevated levels of the latter, thereby inducing leukocyte-endothelial cell interactions. Superoxide dismutase prevented L-NAME-induced leukocyte adhesion,[23] suggesting that removal of nitric oxide may lead to increased superoxide production. These observations have recently been confirmed[24,41] and extended by others; both catalase and desferrioxamine prevented the rise in leukocyte adhesion associated with L-NAME implicating a role for hydrogen peroxide and iron (or iron-catalyzed oxidants) respectively. In addition to preventing leukocyte adhesion, superoxide dismutase also reduced both P-selectin expression and increased leukocyte rolling associated with L-NAME.[31] Finally, the mast cell degranulation associated with L-NAME was also reduced by superoxide dismutase.[23] These data suggest that inhibition of NO synthesis increases levels of superoxide causing a cascade of events including mast cell degranulation, P-selectin expression and leukocyte rolling, adhesion and emigration (summarized in Fig. 2.7).

Detection of oxidative stress is difficult in vivo and quantification virtually impossible. Nevertheless, at least two groups have recently used novel probes that detect intracellular oxidative stress in vivo to demonstrate that L-NAME induces oxidant production.[24,41] The fluorochromes carboxydichlorofluorescein (CDCF) or dihydrorhodamine 123 (DHR) enter cells in a nonfluorescent form, but during increased oxidant production they become oxidized and emit fluorescent light. Suematsu et al[24] demonstrated that superfusion of the rat mesentery with L-NAME caused a rapid increase in oxidant levels within venules, arterioles and tissue mast cells. The venular oxidative impact with L-NAME was equivalent to the same degree of signal invoked by 880 µM tert-butyl hydroperoxide suggesting a significant increase in oxidant production after NO synthesis inhibition. Kurose et al[41] also noted a substan-

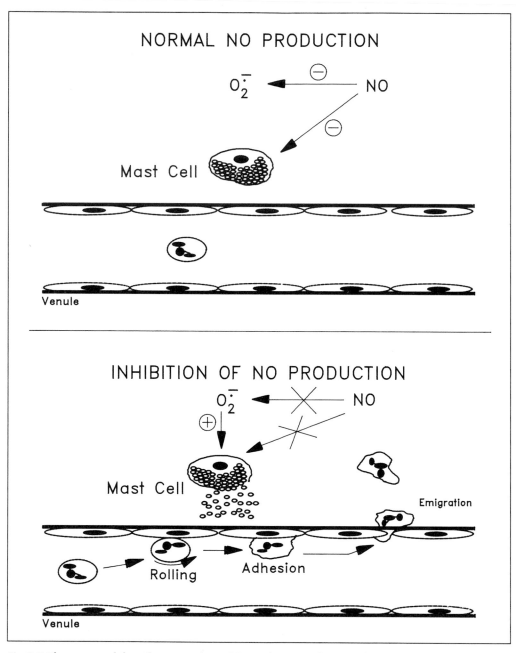

Fig. 2.7. The top panel describes normal conditions wherein endogenously produced NO continuously suppresses mast cell reactivity either directly or by removing endogenously produced superoxide. This may be an autocrine mechanism (both NO and O_2^- produced by mast cells) but both free radicals may be derived from other sources such as the endothelium. When NO synthesis is inhibited, superoxide levels may reach the mast cell, cause degranulation and recruit leukocyte rolling, adhesion and emigration.

tial increase in DHR oxidation following the exposure of tissue to L-NAME. This observation was inhibitable by both catalase and DMTU suggesting a role for hydrogen peroxide or a secondarily-derived oxidant.

Although leukocytes are a primary source of oxidants, the initial increased oxidative stress with L-NAME was seemingly independent of these cells inasmuch as the oxidative stress could be detected prior to leukocyte infiltration.[24,41] Secondly, the oxidative stress occurred at sites distant from adhering leukocytes (arterioles, mast cells). Thirdly, leukocyte-induced oxidative stress demonstrates a different spatial distribution pattern of oxidation than L-NAME.[24] Finally, inhibition of leukocyte adhesion did not affect the initial oxidative stress (first 30 mins) associated with L-NAME further supporting the view that the generation of oxidants in vivo almost certainly precedes leukocyte adhesion.[41] In fact, based on aforementioned data, the increased oxidative stress may be a prerequisite for subsequent leukocyte rolling and adhesion.

Niu et al[50] have recently loaded human umbilical vein endothelial cells with 2',7'dichlorodihydrofluorescein diacetate, another fluorescent probe sensitive to oxidative stress. Interestingly, oxidative stress was detected within 30 mins of L-NAME exposure and continued to rise for the next 2-3 hrs. The oxidative response was equivalent to exposure of endothelial cells to approximately 800 nM hydrogen peroxide; levels that were approximately 1000-fold lower than results in vivo (assuming similar oxidative potential of hydrogen peroxide and *tert*-butyl hydroperoxide). There were other profound differences between the in vivo and in vitro data. The increased oxidative stress was not inhibited by large cell-impermeable antioxidants, including superoxide dismutase and catalase, which likely scavenge extracellular oxidants. By contrast, the oxidative stress was inhibited by sodium azide, the inhibitor of mitochondrial respiration, and by small lipophilic antioxidants that were likely to gain access to the intracellular compartment. Although it is well known that intracellular levels of SOD are quite high (see chapter 1), these data raise the possibility that nitric oxide might behave as an antioxidant within endothelial cells. By contrast the effectiveness of SOD and catalase in vivo to prevent L-NAME-associated responses may suggest that there is an extracellular component to the oxidative stress in the rat mesentery. Oxidative stress in vivo has since been detected in other tissues following L-NAME exposure including the liver.[51] However, the source of the oxidants in these studies remains unknown.

Another difference between the in vivo and in vitro data is the length of time required before leukocyte adhesion became apparent in the in vitro system. Prolonged exposure (2-4 hrs) of endothelial cells to inhibitors of NO synthesis induced leukocyte adhesion, whereas shorter time points were unremarkable in their response to L-NAME.[50] The lack of adhesion at 1 hr despite elevated oxidative stress suggests that oxidant flux may have had to reach critical levels before endothelial cell adhesivity was induced. Indeed, the antioxidants and azide that prevented the oxidative stress in endothelial cells in vitro also reduced leukocyte adhesion but only if they were present throughout the entire 4-hr experiment. If the antioxidants were added after 3 hrs of L-NAME exposure, leukocyte adhesion was still evident; suggesting perhaps that as NO is diminished with L-NAME, mitochondrial-derived oxidants reach critical levels to induce leukocyte adhesion to the endothelium. It would be interesting to add mast cells to this in vitro system to determine whether the L-NAME-treated endothelium could produce oxidants that subsequently activate the mast cell. Under these conditions, one might anticipate earlier leukocyte adhesion (within the first 60 mins) that is inhibitable by extracellular antioxidants (SOD and/or catalase).

The L-NAME-induced adhesive interaction in vitro was further characterized, and the oxidant-induced leukocyte adhesion was shown to involve PAF and CD11/CD18 and to be inhibited by NO donors. Recently, NO donors have been shown to inhibit PAF synthesis by thrombin-activated endothelial cells.[52] Therefore, in addition to potentially acting as a superoxide scavenger, NO may also affect PAF production within endothelial cells and thereby modulate leukocyte-endothelial cell interactions. PAF has been invoked as an important activating molecule that is likely to induce a rolling leukocyte to adhere.[53] Therefore, if NO donors can prevent the synthesis of PAF they could conceivably interfere with the activation of rolling leukocytes and prevent the cascade of events that result in leukocyte recruitment.

Gaboury et al[54] reported that leukocyte adhesion induced by PAF superfusion of the rat mesentery was inhibited by 50% with a nitric oxide donor. In that study, the nitric oxide donor was as effective as an antioxidant (50%), and the authors concluded that nitric oxide could conceivably scavenge PAF-associated superoxide and thereby prevent leukocyte adhesion. Indeed, a superoxide generator that induced leukocyte adhesion and was entirely dependent upon superoxide (inhibitable by SOD) was also completely inhibited by an NO donor. On the other hand, LTB_4 which elicited an SOD-insensitive leukocyte adhesion was not affected by an NO donor. These data imply that 1) NO donors may act as antioxidants to prevent leukocyte adhesion and 2) that not all leukocyte adhesion can be prevented with NO donors.

SUMMARY

In conclusion, there is a growing body of evidence that NO may be a homeostatic regulator of leukocyte adhesion in the microcirculation. Based on this observation and the fact that a hallmark feature of inflammation is the infiltration of leukocytes, our next challenge will be to determine whether exogenous NO (perhaps NO donors) could conceivably reduce the inflammatory process by inhibiting leukocyte infiltration. This concept is presently being tested in ischemia/reperfusion (chapter 5), atherosclerosis (chapter 6) and in one case as a new component of an old drug (nonsteroidal anti-inflammatory drugs) with the untoward side effect of inappropriate leukocyte adhesion (chapter 7). Many questions remain unanswered; the most important of which may be how NO per se inhibits leukocyte adhesion in the microcirculation. This will require more sophisticated approaches including the ability to measure NO levels while simultaneously observing leukocyte behavior online in the microcirculation.

REFERENCES

1. Geng J-G, Bevilacqua MP, Moore KL et al. Rapid neutrophil adhesion to activated endothelium mediated by GMP-140. Nature 1990; 343:757-760.
2. Lawrence MB, Springer TA. Leukocytes roll on a selectin at physiologic flow rates: distinction from and prerequisite for adhesion through integrins. Cell 1991; 65:859-873.
3. Jones DA, Abbassi O, McIntire LV et al. P-selectin mediates neutrophil rolling on histamine-stimulated endothelial cells. Biophysical Journal 1993; 65:1560-1569.
4. Asako H, Kurose I, Wolf R et al. Role of H1 receptors and P-selectin in histamine-induced leukocyte rolling and adhesion in postcapillary venules. J Clin Invest 1994; 93:1508-1515.

5. Kubes P, Kanwar S. Histamine induces leukocyte rolling in post-capillary venules: A P-selectin-mediated event. J Immunol 1994; 152:3570-3577.

6. Bevilacqua MP, Pober JS, Mendrick DL et al. Identification of an inducible endothelial-leukocyte adhesion molecule. Proc Natl Acad Sci USA 1987; 84:9238-9242.

7. Abbassi O, Kishimoto TK, McIntire LV et al. E-selectin supports neutrophil rolling in vitro under conditions of flow. J Clin Invest 1993; 92:2719-2730.

8. Olofsson AM, Arfors KE, Ranezani L et al. E-selectin mediates leukocyte rolling in interleukin-1-treated rabbit mesentery venules. Blood 1994; 84:2749-2758.

9. Ley K, Gaehtgens P, Fennie C et al. Lectin-like cell adhesion molecule 1 mediates leukocyte rolling in mesenteric venules in vivo. Blood 1991; 77:2553-2555.

10. Von Andrian UH, Chambers JD, McEvoy LM et al. Two-step model of leukocyte-endothelial cell interaction in inflammation: distinct roles for LECAM-1 and the leukocyte B_2 integrins in vivo. Proc Natl Acad Sci USA 1991; 88:7538-7542.

11. Lawrence MB, Bainton DF, Springer TA. Neutrophil tethering to and rolling on E-selectin are separable by requirement for L-selectin. Immunity 1994; 1:137-145.

12. Kishimoto TK, Anderson DC. The role of integrins in inflammation. In: Gallin JI, Goldstein IM, Snyderman R, eds. Inflammation: Basic Principles and Clinical Correlates. New York, Raven Press, Ltd, 1992:353-406.

13. Arfors KE, Lundberg C, Lindbom L et al. A monoclonal antibody to the membrane glycoprotein CD18 inhibits polymorphonuclear leukocytes accumulation and plasma leakage in vivo. Blood 1987; 69:338-340.

14. Springer TA. Adhesion receptors of the immune system. Nature 1990; 346:425-434.

15. Bevilacqua MP, Nelson RM. Selectins. J Clin Invest 1993; 91:379-387.

16. Springer TA. Traffic signals of lymphocyte recirculation and leukocyte emigration: The multistep paradigm. Cell 1994; 76:301-314.

17. Furchgott RF, Zawadzki JV. The obligatory role of endothelial cells in the relaxation of arterial smooth muscle by acetylcholine. Nature 1980; 288:373-376.

18. McCall T, Whittle BJR, Boughton-Smith NK et al. Inhibition of FMLP-induced aggregation of rabbit neutrophils by nitric oxide. Br J Pharmacol 1988; 95:517P (abstr.).

19. Kubes P, Suzuki M, Granger DN. Nitric oxide: An endogenous modulator of leukocyte adhesion. Proc Natl Acad Sci USA 1991; 88:4651-4655.

20. Kurose I, Kubes P, Wolf R et al. Inhibition of nitric oxide production: Mechanisms of vascular albumin leakage. Circ Res 1993; 73:164-171.

21. Kubes P, Granger DN. Nitric oxide modulates microvascular permeability. Am J Physiol 1992; 262:H611-H615.

22. Kurose I, Wolf R, Granger DN. Dimethyl-l-arginine, an endogenous inhibitor of nitric oxide (NO) synthesis, elicits an inflammatory response in the rat mesenteric microcirculation. Gastroenterology 1994; 106:A245.

23. Kubes P, Kanwar S, Niu X-F et al. Nitric oxide synthesis inhibition induces leukocyte adhesion via superoxide and mast cells. FASEB J 1993; 7:1293-1299.

24. Suematsu M, Tamatani T, Delano FA et al. Microvascular oxidative stress preceding leukocyte activation elicited by in vivo nitric oxide suppression. Am J Physiol 1994; 266:H2410-H2415.

25. May GR, Crook P, Moore PK et al. The role of nitric oxide as an endogenous regulator of platelet and neutrophil activation within the pulmonary circulation of the rabbit. Br J Pharmacol 1991; 102:759-763.

26. Ma X-L, Weyrich AS, Lefer DJ et al. Diminished basal nitric oxide release after myocardial ischemia and reperfusion promotes neutrophil adherence to coronary endothelium. Circ Res 1993; 72:403-412.

27. Akimitsu T, Korthuis RJ. Leukocyte adherence (LA) induced by inhibition of nitric oxide (NO) production in skeletal muscle. FASEB J 1994; 8:A1032.

28. Nishida J, McCuskey RS, McDonnell D et al. Protective role of nitric oxide in hepatic microcirculatory dysfunction during endotoxemia. Am J Physiol 1994; 267:G1135-G1141.

29. Lopez-Belmonte J, Whittle BJR, Moncada S. Aminoguanidine promotes vasoconstriction and leukocyte adherence in rat mesentery through inhibition of constitutive nitric oxide synthase. Gastroenterology 1995; 108:A302 (abstr.)

30. Tonnesen MG. Neutrophil-endothelial cell interactions: mechanisms of neutrophil adherence to vascular endothelium. J Invest Dermatol 1989; 93:53s-58s.

31. Davenpeck KL, Gauthier TW, Lefer AM. Inhibition of endothelial-derived nitric oxide promotes P-selectin expression and actions in the rat microcirculation. Gastroenterology 1994; 107:1050-1058.

32. Rosen P, Schwippert P, Kaufman B et al. Expression of adhesion molecules on the surface of activated platelets is diminished by PGI2-analogues and an NO (EDRF)-donor: a comparison between platelets of healthy subjects and diabetic subjects. Platelets 1994; 11:42-57.

33. Salvemini D, Masini E, Pistelli A et al. Nitric oxide: A regulatory mediator of mast cell reactivity. J Cardiovasc Pharmacol 1991; 17 (Suppl 3):S258-S264.

34. Hogaboam CM, Bissonnette EY, Befus AD et al. Modulation of rat mast cell reactivity by IL-1 beta. Divergent effects on nitric oxide and platelet-activating factor release. J Immunol 1993; 151:3767-3774.

35. Galli SJ. New concepts about the mast cell. N Engl J Med 1993; 328:257-265.
36. Gaboury JP, Kubes P. Mast cell degranulation causes granulocyte rolling and adhesion in venules. Gastroenterology 1994; 106:A1027.
37. Klein LM, Lavker RM, Matis WL et al. Degranulation of human mast cells induces an endothelial antigen central to leukocyte adhesion. Proc Natl Acad Sci USA 1989; 86:8972-8976.
38. Gaboury JP, Johnston B, Niu X-F et al. Mechanisms underlying acute mast cell-induced leukocyte rolling and adhesion in vivo. J Immunol 1995; 154:804-813.
39. Crowe SE, Perdue MH. Gastrointestinal food hypersensitivity: Basic mechanisms of pathophysiology. Gastroenterology 1992; 103:1075-1095.
40. Kanwar S, Wallace JL, Befus D et al. Nitric oxide synthesis inhibition increases epithelial permeability via mast cells. Am J Physiol 1994; 266:G222-G229.
41. Kurose I, Wolf R, Grisham MB et al. Microvascular responses to inhibition of nitric oxide production: role of active oxidants. Circ Res 1995; 76:30-39.
42. Arndt H, Russell JM, Kurose I et al. Mediators of leukocyte adhesion in rat mesenteric venules elicited by inhibition of nitric oxide synthesis. Gastroenterology 1993; 105:675-680.
43. Wallace JL. PAF as a mediator of gastrointestinal damage. In: Saito K, Hanahan DJ, eds. Platelet-Activating Factor and Diseases. International Medical Publishers, 1989:153-186.
44. Rubanyi GM, Vanhoutte PM. Superoxide anions and hyperoxia inactivate endothelium-derived relaxing factor. Am J Physiol 1986; 250:H822-H827.
45. Del Maestro RF, Planker M, Arfors KE. Evidence for the participation of superoxide anion radical in altering the adhesive interaction between granulocytes and endothelium, in vivo. Int J Microcirc Clin Exp 1982; 1:105-120.
46. Gaboury J, Anderson DC, Kubes P. Molecular mechanisms involved in superoxide-induced leukocyte-endothelial cell interactions in vivo. Am J Physiol 1994; 266:H637-H642.
47. Granger DN. Role of xanthine oxidase and granulocytes in ischemia-reperfusion injury. Am J Physiol 1988; 255:H1269-H1275.
48. Suzuki M, Inauen W, Kvietys PR et al. Superoxide mediates reperfusion-induced leukocyte-endothelial cell interactions. Am J Physiol 1989; 257:H1740-H1745.
49. Cross CE. Oxygen radicals and human disease. Ann Int Med 1987; 107:526-545.
50. Niu X-F, Smith CW, Kubes P. Intracellular oxidative stress induced by nitric oxide synthesis inhibition increases endothelial cell adhesion to neutrophils. Circ Res 1994; 74:1133-1140.

51. Bautista AP, Spitzer JJ. Inhibition of nitric oxide formation in vivo enhances superoxide release by the perfused liver. Am J Physiol 1994; 266:G783-G788.
52. Heller R, Bussolino F, Ghigo D et al. Nitrovasodilators inhibit thrombin-induced platelet-activating factor synthesis in human endothelial cells. Biochem Pharmacol 1992; 44:223-229.
53. Zimmerman GA, Prescott SM, McIntyre TM. Endothelial cell interactions with granulocytes: tethering and signaling molecules. Immunology Today 1992; 13:93-99.
54. Gaboury J, Woodman RC, Granger DN et al. Nitric oxide prevents leukocyte adherence: role of superoxide. Am J Physiol 1993; 265:H862-H867.

BIOLOGICAL SIGNIFICANCE OF NITRIC OXIDE IN PLATELET FUNCTION

Marek W. Radomski and Eduardo Salas

INTRODUCTION

Platelets play an important role in vascular homeostasis. Physiologic platelet reactions help to restore the compromised integrity of vessel wall and maintain homeostasis. The platelet reactions are composed of highly sophisticated activation responses and often function in a cascade-like manner. Under normal conditions the platelet activation is well-controlled at several biochemical check-points. This physiologic platelet response is known as platelet hemostasis. However, vascular disorders may impair the regulatory check-points and precipitate uncontrolled reactions platelet thrombosis.

The biosynthesis of nitric oxide (NO) in the vascular system provides a simple but efficient regulatory mechanism to support hemostasis and prevent thrombosis. In this chapter, we will review evidence for the importance of NO as a hemostatic molecule. We will also describe how the changes in the generation, actions and metabolism of NO may affect the pathogenesis of vascular diseases. Finally, we will present the pharmacologic approaches that can modulate the clinical course of vascular pathologies.

Nitric Oxide: A Modulator of Cell-Cell Interactions in the Microcirculation, edited by Paul Kubes. © 1995 R.G. Landes Company.

PHYSIOLOGIC PLATELET REACTIONS

Platelets are small anucleate blood elements that are formed by fragmentation of large mother cells, megakaryocytes. The presence of plasma granules is the most prominent feature of platelet ultra-structure. The granules contain activating and proliferating agents whose release is vital for platelet responses. Pulsatile blood flow and the shear rate are major determinants of platelet behavior in vivo.[1] The shear rate is largely responsible for the tendency of suspended particles (blood elements) to move towards the center of the flowing stream. If platelets were the only formed element in the blood, they would occupy the axial stream and be unlikely to interact with the vascular wall. However, in the whole blood, the more numerous and larger erythrocytes occupy the axial stream and force platelets to assume the position close to endothelial cells. Thus, under physiologic conditions, platelets remain in a close contact with the endothelium, yet these interactions do not cause activation. However, when a blood vessel is damaged and the endothelium is disrupted, platelets are quickly recruited from blood to the place of injury where they form an occlusive plug. The biological meaning of these reactions is "to seal rents" in the vascular system. Since the sealing processes are very complex it is convenient to separate them into: adhesion, the ability of a platelet to make contact and spread on a natural or foreign surface; aggregation, the ability of platelets to attach to other platelets; and platelet coagulant activity, the ability of platelets to accelerate the activation of the coagulation cascade that leads to the formation of thrombin and reinforcement with fibrin of platelet aggregates. During activation platelets undergo startling metamorphosis; they change shape from discoid to ameboid with multiple, sometimes long cytoplasmic projections. What mechanisms are responsible for these elaborated changes?

MOLECULAR BASIS OF PLATELET ACTIVATION

ADHESION

The biological signal for initiating platelet adhesion is delivered by the exposure of adhesive portions of vessel wall (subendothelium, media, adventitia) that are normally concealed from the blood by an intact endothelial monolayer.[2] Platelets make contact with these adhesions using specific receptors (integrins or other) that anchor them to the components of the subendothelial matrix.

AGGREGATION

Platelets possess a number of receptors that are recognized by soluble factors circulating in plasma such as epinephrine, ADP, vasopressin, serotonin and thrombin. These factors act in concert to amplify the signal initiated by adhesion to the subendothelial matrix. The amplification of the platelet membrane signaling system involves activation of lipid-metabolizing enzymes including phospholipases A_2 and C. Phospholipase A_2 liberates arachidonic acid from cellular lipids which is further converted by cyclooxygenase to cyclic endoperoxides and by thromboxane synthase to thromboxane A_2 (TXA_2) which potentiate platelet activation. Phospholipase C converts membrane phosphatidylinositol diphosphate into diacylglycerol and inositol triphosphate that activate protein kinase C and the release of Ca^{2+} from the dense granules respectively. Protein kinase C phosphorylates substrate proteins while Ca^{2+} initiates a number of platelet reactions. These biochemical changes are accompanied by a dramatic reorganization of platelet cytoskeleton proteins leading to the change of platelet shape. At the same time platelet granules are grouped and secreted outside by the force generated by platelet contractile proteins (internal contraction). The secretion of Ca^{2+} and other granule constituents (the platelet release reaction) is clearly linked to platelet aggregation. The platelet releasate contains a number of agents that support aggregation and act via amplification of the release reaction (ADP and TXA_2) or by stabilizing loose bonds between adjacent platelets (fibrinogen and von Willebrand factor). The stabilization of platelet-platelet bridges is achieved by the conversion of platelet glycoprotein integrin receptor IIb/IIIa to the active conformation that binds tightly to fibrinogen. Finally, platelet secretion may also facilitate the formation of platelet-leukocyte aggregates by allowing α-granule protein P-selectin to be translocated to the platelet surface membrane and bind to the leukocyte ligand.[3]

PHYSIOLOGIC REGULATION OF PLATELET ACTIVATION

The process of platelet activation is under tight control. At least three regulatory systems have been described that regulate platelet reactivity. These include cyclooxygenase and lipoxygenase metabolites (particularly prostacyclin and 13-hydroxyoctadecadienoic acid or 13-HODE), ecto-nucleotidase ATP-diphosphohydrolase and NO.

PROSTACYCLIN AND 13-HODE

Prostacyclin is an eicosanoid synthesized from arachidonic acid by the sequential actions of cyclooxygenase and prostacyclin synthase enzymes.[4] Albeit labile prostacyclin is a potent vasodilator and inhibitor of platelet activation. Acting at very low concentrations (nM), prostacyclin inhibits aggregation and stimulates disaggregation of preformed platelet aggregates. Interestingly, prostacyclin appears to be a weak inhibitor of platelet adhesion.[5] These platelet-inhibitory effects of prostacyclin are mediated by stimulation of specific receptors linked to the adenylate cyclase system leading to an increase in cAMP levels. Cyclic AMP-mediated responses result in the phosphorylation of specific platelet proteins followed by down-regulation of several stages of the platelet transduction mechanism including inhibition of intraplatelet Ca^{2+} release and activation of platelet receptors.[6]

The metabolism of linoleic acid through the lipoxygenase pathway leads to the generation of 13-HODE.[7] It has been suggested that 13-HODE is an intracellular regulator of vessel wall adhesiveness and a modulator of integrin receptor expression.

ECTO-NUCLEOTIDASES

These are platelet-regulatory enzymes that metabolize ADP to AMP and adenosine, thereby limiting platelet recruitment and reactivity.[8] It appears that adenosine inhibits platelet function via cAMP-dependent mechanisms.

NITRIC OXIDE

A HISTORIC PERSPECTIVE

The current awareness of biological significance of NO as a regulator of platelet function has evolved from the pioneer findings by Furchgott and Zawadzki[9] who demonstrated that the vascular endothelium, when stimulated with acetylcholine, released an unstable substance (endothelium-derived relaxing factor, EDRF) that caused vasorelaxation of rat aorta rings. The search for the chemical identity of this factor led to the proposal that the free radical NO gas is EDRF[10,11] and culminated with the description of NO biosynthesis, L-arginine to NO pathway.[12,13] This pathway has been shown to be of crucial importance for the regulation of homeostasis in cardiovascular, central and peripheral, respiratory, gastrointestinal, genitourinary, endocrine and immune systems.[14]

In the vascular system, the abluminal release of NO by endothelial cells results in the vasodilator tone, increased blood flow, decreased resistance and increased conductivity of the vessel wall.[14] In this chapter, we will focus on the generation, release and actions of NO in the lumen of arteries and veins. The actions of NO inside the vascular lumen are important for the preservation of blood fluidity. It is noteworthy that the conductile and fluid properties of vessel wall are mutually dependent in maintaining vascular homeostasis.

NITRIC OXIDE SYNTHASES

General characteristic

There are at least three separate genes encoding the nitric oxide synthase enzyme family of proteins (NOS). These are nNOS from neurons, eNOS from endothelium and iNOS, an isoform inducible in many cell types.[15] The studies on the chromosomal localization of human NOS have assigned eNOS to 7, nNOS to 12 and iNOS to 17. The isoforms of NOS belong to the family of the cytochrome P450 reductase enzyme. The native NOS isoenzyme appear to be homodimers localized both in the soluble and particulate fractions of the cell with molecular weight ranging from 130 to 160 per subunit. Consensus binding sites for FAD, FMN, NADPH and calmodulin are conserved for all isoforms. Binding sites for L-arginine, H_4biopterin, heme and molecular oxygen have also been postulated. Some NOS (e.g. nNOS and eNOS) are constitutive in the sense that their activation does not require new enzyme protein synthesis and usually depends on increased intracellular Ca^{2+} levels to stimulate the binding of calmodulin to its site and NOS activation.[16] Other NOS (e.g. iNOS) are inducible, i.e., their activation requires new protein synthesis. Endotoxin and cytokines often induce the expression of iNOS. A distinct feature of iNOS is also its independence of Ca^{2+} transients that probably results from the observation that iNOS contains calmodulin as a tightly bound subunit at the low level of Ca^{2+} found in resting cells. The iNOS, although functionally Ca^{2+}-independent, may be regulated at transcriptional, translational and posttranslational levels.[17]

The platelet NOS

In 1990 we found that platelets generate NO via a NOS-dependent pathway.[18,19] The rationale for this work was derived from the observations that the activity of soluble guanylyl cyclase (GC-S) and the levels of cyclic GMP (cGMP) were elevated during platelet aggregation by aggregating agents such as collagen and arachidonic acid. These observations prompted the hypothesis that cGMP may be one of the mediators of platelet activation that acts by opposing the platelet-inhibitory activity of cAMP.[20] However, in the early eighties the work by Ignarro and colleagues clearly demonstrated that the biological actions of cGMP inhibited platelet aggregation.[21] In order to resolve this controversy, we hypothesized that aggregated platelets express NOS and that NO forms an autocrine regulatory mechanism that down-regulates aggregation via a cGMP-dependent mechanism.[18] This hypothesis was supported by some earlier work that demonstrated that L-arginine, the substrate for NOS, inhibited platelet aggregation both ex vivo[22] and in vitro.[23] Using platelet aggregometry we confirmed that L-arginine was an inhibitor of platelet aggregation and found that inhibition of NOS potentiated aggregation induced by different aggregating agents in vitro.[19] We also found that cytosolic fractions obtained from human platelet homogenates constitutively expressed Ca^{2+}-dependent NOS activity which was functionally linked to the activity of GC-S.[18] The presence of NOS was subsequently confirmed in human[24-33] and described in rabbit,[34] porcine,[35] canine[36] and rat platelets.[37] Recently, two groups have isolated and purified to homogeneity NOS from the cytosolic fraction of human platelets.[28,33] The native proteins appeared to be dimeric, Ca^{2+}-, calmodulin-, NADPH-, FAD-, H_4biopterin- and L-arginine-dependent, and their activities were sensitive to the inhibition with inhibitors of NOS (Table 3.1). Interestingly, the reported molecular weights of identified subunits of NOS were 80 and 130 kDa, respectively.[28,33] Whether this observation reflects some methodological differences in the isolation of NOS or 80 kDa protein is a NOS with a novel molecular weight remains to be established. All these data clearly indicate that platelets constitutively express NOS with the biochemical profile similar to nNOS and eNOS.

The expression of iNOS in cells requires de novo protein synthesis. Platelets are anucleate and acquire most of their proteins by transfer from parent cells, megakaryocytes.[38] From these rea-

Table 3.1. Characteristics of nitric oxide synthase (NOS) and the soluble guanylate cyclase (GC-S) in platelets.

	NOS	GC-S
Protein	homodimer	heterodimer $\alpha_1\beta_1$
	80-130 kDa	73 and 70 kDa
Cofactors	heme?	heme present
	Ca^{2+}	Ca^{2+}
	calmodulin	Mg^{2+}
	NADPH	
	H_4biopterin	
Substrate	L-arginine	GTP
Products	NO	cGMP
	L-citrulline	
Inhibitors of enzyme activity or action	Arginine analogs	ODQ
	methylene blue	LY83583
	hemoglobin	methylene blue

sons we suggested that most of the platelet iNOS activity is likely to be derived from megakaryocytes which can express both constitutive and inducible NOS.[39] However, recent work has established the presence of platelet-specific mRNAs in platelets and showed that they do synthesize proteins.[40] In addition, incubation of platelets with some cytokines resulted in the expression of iNOS activity.[33] Thus, the platelet iNOS may be both of platelet and megakaryocyte origin.

Interestingly, it appears that platelets themselves can modulate the expression of iNOS. Indeed, platelet-derived growth factor (PDGF) released from platelet granules by collagen has been shown to down-regulate interleukin-1β (IL-1β)-induced expression of iNOS in rat vascular smooth muscle cells.[41] In contrast, the exposure of cells to platelet surface membranes could promote the expression of iNOS since activated platelets show membrane-bound IL-1β capable of cytokine induction in endothelial cells.[42]

The endothelial NOS

The endothelial cells express also a constitutive isoform of NOS (eNOS). The eNOS from human umbilical vein and bovine aorta endothelial cells have been cloned.[43,44] This is a Ca^{2+}-, NADPH-, flavin- and H_4biopterin-dependent enzyme. The N-terminal myristoylation of eNOS is a unique feature amongst the NOS family and permits this isoform to be compartmentalized and associated with the particulate fraction of the endothelium.[16]

The exposure of endothelial cells to endotoxin and cytokines results in the expression of iNOS activity.[45] The iNOS from endothelial cells has not been cloned yet, however, Northern and Western blots analyses have confirmed the cytokine-induced expression of both protein and mRNA for iNOS in rat vascular endothelial cells.[46] The biochemical characteristic showed that this isoform, similarly to that from macrophages, requires NADPH, H_4biopterin and L-arginine and is Ca^{2+}-independent.[45,47] The expression of endothelial iNOS is inhibited by transforming growth factor (TGF-β), inhibitors of protein synthesis and by glucocorticoids.[45,46] TGF-β and glucocorticoids may block iNOS gene expression at the transcriptional level.[46]

NITRIC OXIDE SYNTHASE REACTION

The isoforms of NOS utilize the guanido nitrogen atom of L-arginine and incorporate molecular oxygen to generate NO and L-citrulline.

NOS uses NADPH as an electron donor to catalyze the hydroxylation of L-arginine to N-hydroxy-L-arginine, an intermediate in the pathway. N-hydroxy-L-arginine is then oxidized to yield NO and L-citrulline. Interestingly, NOS is the only known mammalian enzyme catalyzing a hydroxylation reaction and NADPH reduction within the same protein. Some arginine analogs including N^G-monomethyl-L-arginine (L-NMMA) inhibit both parts of the NOS reaction.[15-17]

METABOLIC FATE OF NO

Both physicochemical properties and the microenvironment determine the metabolic fate of NO. In principle, there are three redox states of NO: free radical (NO), nitrosonium (NO^+) and nitroxide (NO^-). NO is a paramagnetic and diffusible molecule. The diffusivity of NO closely resembles that of oxygen[48] and at a given temperature is smaller in proteins and membranes than in water. NO is likely to accumulate in lipids since its partitioning is greater in hydrocarbon than in water. The presence of endogenous higher oxides of oxygen, nitrite and nitrate suggests that the reaction of NO with molecular oxygen may be one of the major determinants of its metabolism. However, the experiments using recently developed porphyrinic electrochemical microsensors indicate that molecular oxygen plays a minor role in the direct oxidation process of NO in biological systems.[49] Indeed, it appears that NO reacts with oxygen to form nitrite only at high, nonphysiologic concentrations. In contrast, NO reacts at a near diffusion-limited rate with superoxide to generate peroxynitrite (ONOO$^-$).[50]

NO$^+$ may be involved in the reactions of S-nitrosylation to form S-nitrosothiols. There is some evidence that S-nitrosylation of endogenous thiols such as glutathione and albumin results in the formation of S-nitrosothiols and prolongation of biological half-life of NO. S-nitrosylation of reactive thiols can also modify the activity of receptor and enzyme proteins.[51] Interestingly, ONOO$^-$ can also result in S-nitrosylation and generation of S-nitrosothiols.[52]

Nitroxide has been proposed as a physiologically relevant form of NO,[53] however it is possible that most of biological actions of NO$^-$ (if proved to be generated in vivo) are indirect and result from its conversion to NO.[54]

INTERACTIONS OF NO WITH BIOMOLECULES

Nitric oxide has an affinity to heme in heme proteins including GC-S.[55] Recently, the sequestration of NO gas by subcellular fractions of vascular smooth muscle and platelets has been examined using the chemiluminescence-headspace gas technique.[56] This study has demonstrated that NO is sequestered preferentially by subcellular fractions of these cells that contain GC-S activity and that the sequestration of NO in these fractions stimulates the catalytic activity of GC-S.

The isoforms of GC-S are heterodimeric proteins consisting of two α and β subunits. The human platelet GC-S has been characterized using polyclonal antibodies raised against synthetic peptides corresponding to different subunits of the lung isoform.[57] The study has showed the platelet enzyme to be α1β1 hemoprotein exhibiting molecular masses of 73 and 70 kDa respectively (Table 3.1). The presence of heme as a prosthetic group and the simultaneous expression of both subunits are required for its catalytic activity.[58] The reaction leads to the conversion of magnesium guanosine 5'-triphosphate to guanosine 3',5'-monophosphate (cGMP). Although the evidence for its inhibitory action on platelet responses is overwhelming, the levels of this inhibition are still being elucidated

(Fig. 3.1). Elevated concentrations of cGMP activate cGMP-dependent protein kinase and the phosphorylation of various target proteins.[59] Among proteins phosphorylated in response to cGMP the best characterized is 46/50 kDa vasodilator-stimulated phosphoprotein (VASP).[60] In adhering platelets VASP is associated with actin filaments and focal contact areas, i.e., transmembrane junctions between microfilaments and the extracellular matrix.[61] In particular, the association of VASP with the platelet cytoskeleton may be of importance for its inhibitory effect on the fibrinogen receptor.[62]

Cyclic GMP-induced protein phosphorylation may also be involved in the uptake of serotonin by platelets.[30]

Cyclic GMP decreases basal and stimulated concentrations of intracellular Ca^{2+}.[63,64] A number of Ca^{2+} handling systems have been identified in platelets including receptor operated channels, passive leak, Ca^{2+}-ATPase extrusion pump, the Na^+/Ca^{2+} exchanger, Ca^{2+}-accumulating ATPase pump of the dense tubular membrane (an intraplatelet membrane Ca^{2+} store) and passive leakage and receptor operated Ca^{2+} channels in the dense tubular membrane. In principle, all these processes could be affected by cGMP. It has been shown that cGMP increases the activity of Ca^{2+}-ATPase extrusion pump and leakage across the plasma membrane.[64] This extrusion pump may be, in fact, VASP.[64] In addition, cGMP causes inhibition of Ca^{2+} mobilization from intraplatelet stores including the dense tubular membrane.[63] In contrast to cAMP, cGMP does stimulate the dense tubular Ca^{2+} pump. Consequently, cGMP cannot result in increased dense tubular sequestration of Ca^{2+} whereas

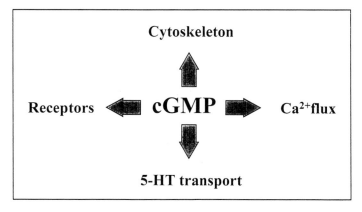

Fig. 3.1. Cyclic GMP as a regulator of platelet function.

cAMP does. Thus, cGMP may be "a better Ca^{2+} antagonist".[64] Under some conditions cGMP by inhibiting cGMP-inhibited cAMP phosphodiesterase may delay the hydrolysis of cAMP and enhance the biological effects of the latter nucleotide.[65] However, the physiologic and pharmacologic relevance of this "cross-talk" between cAMP and cGMP pathways is unclear.[66] Metabolism of membrane phospholipids may also be a target for the action of cGMP. Indeed, the inhibition of both phospholipase C and A$_2$ has been implicated in the mechanism of this action on platelets.[63,67] Finally, cGMP down-regulates the function of some platelet receptors including the fibrinogen receptor IIb/IIIa and P-selectin.[68,69]

The biological actions of cGMP are terminated by cGMP phosphodiesterases as well as by its efflux from platelets.[59,70]

There is also ongoing dispute as to the contribution of cGMP-independent effects in regulation of platelet function by NO (Fig. 3.2). In addition to GC-S, NO may target arachidonic acid cyclooxygenase and 12-lipoxygenase enzymes, some enzymes of mitochondrial respiratory chain, glyceraldehyde-3-phosphate dehydrogenase (GAP-DH) and intracellular thiols such as glutathione. Recent studies have demonstrated that NO is unlikely to interact with heme-containing cyclooxygenase and may selectively inhibit the activity of 12-lipoxygenase, an enzyme that does not contain

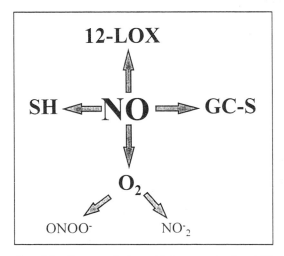

Fig. 3.2. Some of the affinity targets for NO in platelets.
12-LOX: arachidonic acid 12-lipoxygenase enzyme
SH: reactive thiols
GC-S: the soluble guanylyl cyclase enzyme.

heme.[29] Nitric oxide-induced inhibition of mitochondrial enzymes and GAP-DH could interfere with oxidative and glycolytic pathways of ATP formation,[71] although the significance of these effects to platelet function remains to be established. Interestingly, GAP-DH activation may be required for the transport of NO from some S-nitrosothiols to platelets.[72] Studies attempting to investigate respective contributions of cGMP-dependent and cGMP-independent actions to platelet-inhibitory activity of NO have been hampered by the lack of a potent and selective inhibitor of GC-S. We have recently characterized 1H-[1,2,4]oxadiazolo[4,3-a]quinoxalin-1-one (ODQ) as a potent and selective inhibitor of GC-S (Moro MA, Russell RJ, Cellek S, Lizansoain I, Su Y, Darley-Usmar, Radomski MW, Moncada S, unpublished data). Using this compound we have shown that the antiaggregatory effect of NO in vitro could be completely reversed by ODQ and thus is cGMP-dependent.

In addition to NO, the activity of GC-S can be increased by hydrogen peroxide.[73] arachidonic acid hydroperoxides,[74] and carbon monoxide.[75] The significance of these stimuli for the regulation of GC-S in vivo remains to be determined.

Among other possible reactants of NO are: albumin, hemoglobin and thiol-containing cellular receptors and ion channels. S-nitrosylation of albumin led to the generation of S-nitrosoalbumin that could serve as a biological depot for NO.[76] In contrast, red cell hemoglobin is believed to be an NO scavenger that decreases the amount of biologically active NO.[77] The interactions of NO with platelet receptors and ion channels remain to be investigated.

NITRIC OXIDE—PHYSIOLOGIC REGULATOR OF PLATELET FUNCTION

Platelet-derived NO

Early studies using indirect (activation of GC-S) and direct (electrochemical measurement of NO) assays failed to detect the generation of NO by unstimulated platelets.[18,78] Recently, Zhou and colleagues have found that platelets isolated from human blood, in the absence of aggregating agents, can generate NO.[32] This release was partially inhibited by agents known to inhibit platelet activation suggesting that platelets could have been activated during isolation procedures. However, if proven to occur in vivo, the basal production of NO from platelets could generate as much as 20 nmol NO per 1L of blood (based on $1\text{-}3 \times 10^{11}$ platelets in 1L of blood).[32] The platelet NOS becomes activated during platelet adhesion to collagen[79] and aggregation induced by collagen, ADP and arachidonic acid.[18,19,32,78] Nitric oxide generated during these reactions down-regulates the extent of platelet activation. Since this enzyme is strictly Ca^{2+}-dependent and platelet aggregation is associated with an increase in intraplatelet Ca^{2+}, it is possible that this cation controls the activation of the platelet NOS. However, our recent studies using direct electrochemical measurement of NO released from human platelets have shown that NO is released during collagen- but not thrombin-induced aggregation.[78] The reasons for the differential action of collagen and thrombin on NO release are not clear since both aggregating agents are known to mobilize Ca^{2+} from intraplatelet compartments. Since differential Ca^{2+} compartments have been described in human platelets,[80] the activation of the platelet NOS may be linked to a collagen-sensitive but thrombin-insensitive pool. Alternatively, NO generated during platelet aggregation with thrombin is rapidly converted to a metabolite that is not detected by the NO-selective electrochemical system.

Nitric oxide generated by endothelial cells and leukocytes

The amounts of NO available for regulation of platelet function are supplemented by its generation in vascular endothelial cells. Studies with inhibitors of NOS suggest that eNOS generates NO constantly to provide the vasodilator tone.[81] The physiologic stimuli for generation of NO by the endothelium are not yet fully understood, but flow and shear stress seem to stimulate the synthesis of NO via activation of the potassium K_{Ca} channel.[82]

In 1987 it was also shown that cultured and fresh endothelial cells, when stimulated with bradykinin, released NO in quantities sufficient to inhibit platelet adhesion.[5,83,84] Moreover, the coronary and pulmonary vasculature generated NO to inhibit platelet adhesion under constant flow conditions.[85,86]

Platelet aggregation in vitro induced by a variety of agonists as well as by shear stress is inhibited by NO released from fresh or cultured endothelial cells.[77,87-93] This NO also causes disaggregation of preformed platelet aggregates.[87]

Regulation of platelet function by NO in vivo

The animal studies have demonstrated that basal[34,36,94-96] or stimulated (by cholinergic stimuli and substance P) NO results in inhibition of platelet aggregation induced by some aggregating agents or endothelial injury and increases bleeding time.[97-100] In addition, there is luminal release of NO from human vasculature causing increases in intraplatelet cGMP levels.[101] Finally, the administration of NOS inhibitor L-NMMA into healthy volunteers increased platelet aggregation and granule release,[27] whereas L-arginine, the substrate for NO synthesis, led to the inhibition of platelet activation.[102] Thus, a concerted action of cellular NOS regulates platelet activation, causing inhibition of adhesion and aggregation and induction of disaggregation. Both the vasodilator[100] and platelet-inhibitory components[34,36,94-99,101] contribute to the hemostatic action of NO.

The contribution of NO released from neutrophils[103] to the regulation of platelet function in vivo remains to be established. However, it is of interest that the NO-cGMP system also inhibits the adhesion and chemotaxis of stimulated neutrophils.[104,105]

The synthesis and release of a single inhibitor is unlikely to account for regulation of platelet aggregation. We have shown that NO and prostacyclin synergize with each other as inhibitors of platelet aggregation and inducers of disaggregation.[87] In addition, synergistic induction of platelet disaggregation has been demonstrated by combining glyceryl trinitrate (a NO donor), prostaglandin E_1 and tissue plasminogen activator, which act via cGMP, cAMP and plasmin-dependent mechanisms respectively.[106] Furthermore, the inhibition of NO generation in vivo abolishes the antithrombotic activity of aspirin (inhibits the generation of prothrombotic eicosanoids) and ketanserin (antagonizes the actions of serotonin).[34] Thus, it is likely that platelet aggregation in vivo is regulated by the synergistic action of inhibitors of platelet function.

ROLE OF NO IN THE PATHOGENESIS OF VASCULAR DISORDERS

The vasodilator and platelet-regulatory functions of endothelium are impaired during the course of essential hypertension, diabetes and atherosclerotic coronary artery disease,[107] however the pathomechanism of these changes remains unclear. Since oxidative modification of low-density lipoproteins (LDL) plays a key role in atherogenesis, a number of studies have examined the effects of native and oxidized LDL on NO-mediated vascular functions. In most of these studies lipoproteins decreased the bioactivity of NO. Several hypotheses have been proposed to explain these effects of

LDL including inhibition of NOS activity, direct inactivation of NO or changes in NO metabolism.[107-110] In addition, LDL may inhibit arginine uptake into platelets and through this mechanism decrease NOS activity and promote thrombosis.[111] Interestingly, high-density lipoproteins (HDL) decreased platelet function by increasing NOS activity in platelets.[31] Finally, oral administration of L-arginine to hypercholesterolemic rabbits decreased atherosclerosis-induced platelet activation by stimulating NOS in platelets.[112] An impaired NO generation or action may also underlie the pathomechanism of vasospastic and thrombotic changes of essential hypertension,[26,113] diabetes[114] and the coronary artery disease.[107] Indeed, the endogenous NO inhibited microthromboembolism in the ischemic heart, protected myocardium against intracoronary thrombosis and decreased platelet deposition due to the carotid endarterectomy.[115,116]

Interestingly, some researchers reported increased generation of nitrogen oxides by iNOS during atherogenesis that was associated with decreased bioactivity of these compounds.[117] There is evidence that the cellular toxicity associated with the expression of inducible NO is indirect and mediated via formation of $ONOO^-$ from NO and superoxide anion. Peroxynitrite is a potent and highly reactive oxidant which, when formed, can propagate the cascade of oxidation and cell damage.[50] Indeed, it has been demonstrated that $ONOO^-$ may play an important role in oxidant-induced endothelial injury.[118] Recently, we have found that $ONOO^-$ stimulated aggregation of human platelets and reversed inhibition of aggregation by NO and prostacyclin.[52] In addition, the exposure of the isolated perfused coronary vasculature to this oxidant led to a decrease in vascular reactivity to the vasodilators.[119] Interestingly, the effects of $ONOO^-$ on platelets were attenuated in the presence of thiols due to its conversion to S-nitrosothiol(s) which inhibited platelet aggregation.[50] In addition, $ONOO^-$ induced impairment in vascular reactivity that was prevented by exogenous NO or prostacyclin.[119] Thus, the net effect (cytoprotective or cytodestructive) of inducible NO is likely to be dependent on the microenvironment where this molecule is generated, reacts and acts. Therefore, the conversion of NO to oxidants such as $ONOO^-$ may play a role in the pathogenesis of atherogenesis and other vascular pathologies.[52,119,120]

The expression of iNOS has also been implicated in the pathogenesis of septicemia and septic shock. The iNOS is induced following the invasion of gram-negative bacteria and exposure of cells to endotoxin (LPS) and cytokines.[107] LPS- and cytokine-mediated expression of iNOS is likely to have complex repercussions for vascular hemostasis. On one hand, inducible NO acts to promote hemostasis and inhibit thrombosis since the inhibition of its generation by NOS inhibitors greatly potentiated cytokine-stimulated platelet adhesion to cultured human endothelial cells,[121] precipitated renal glomerular thrombosis[122] and exacerbated sepsis-induced renal hypoperfusion.[123] On the other hand, the exposure of endothelial cells to cytokine-induced NO may result in cell toxicity and destruction,[124] and it has been reported that the inhibition of NOS may be beneficial in the treatment of septic shock.[125,126] A partial explanation for this discrepancy may be that the currently available inhibitors of iNOS are not selective and inhibit the activities of other NOS isoenzymes. Indeed, intact generation of constitutive NO may be important to maintain the integrity of the microvasculature during sepsis.[126] Clinical investigations are in progress to address the efficacy and safety of NOS inhibition treatment in this condition.

Bleeding is a well-known complication of uremia that is attributed to the suppression of platelet function by the disease process. Interestingly, L-arginine and some other platelet-inhibitory metabolites of the urea cycle are accumulated in uremia,[127] and this is associated with an increase in TNFα levels.[25] It has been shown that platelets obtained from uremic patients generate more NO than controls so that increased expression and/or activity of NOS may play a role in platelet dysfunction observed in uremia.[25]

Platelets play a role in the pathogenesis of tumor metastasis by increasing the formation of tumor cell-platelet aggregates thus facilitating cancer cell arrest in the micovasculature. We have demonstrated that tumor cell-induced platelet aggregation in vitro is modulated by the ability of tumor cells to generate NO, and this correlated with their propensity for metastasis.[128] Indeed, human colon carcinoma cells isolated from metastases exhibited lower NO activity than cells isolated from the primary tumor. Moreover, the expression of iNOS by murine melanoma cells inversely correlated with their ability to form metastases in vivo.[129] These data suggest that differential synthesis of NO may distinguish between cells of low and high metastatic potential. Interestingly, NOS has been found in some human gynecological malignancies and the highest NOS activities detected in poorly differentiated tumors.[130] Thus, further work is needed to unravel the biological significance of NO generation by cancer cells.

PHARMACOLOGIC REGULATION OF PLATELET FUNCTION BY NO GAS AND NO DONORS DRUGS

Nitric oxide gas is a potent inhibitor of platelet activation in vitro, however its antiplatelet activity is limited by short chemical (sec) and biological (1-4 mins) half-lives.[77] Interestingly, inhalation of NO gas by healthy volunteers and experimental animals resulted in a longer-lasting (appr. 30 mins) inhibition of platelet hemostasis as detected by prolongation of bleeding time.[131] This phenomenon may be explained either by accumulation and subsequent slow release of NO from the lipid part of cell membrane[132] or formation of endogenous NO donors.

The mechanism of pharmacologic activity of NO donors is mostly due to the release of NO.[133] The main advantage of these compounds is their longer than NO gas half-life. The NO donors can be classified according to their ability to release NO in vitro. The first group is comprised of the drugs that require metabolic transformation to release NO.

Organic nitrates are the most prominent members of this class. In vitro, these compounds are poor spontaneous releasers of NO and require the presence of a thiol cofactor (e.g. N-acetylcysteine) for acceleration of this liberation.[134] However, in vivo the release of NO from organic nitrates is greatly enhanced by enzyme(s) which still remain to be identified. These enzymes are present in the vascular tissues but not in platelets.[134] Indeed, organic nitrates are weak inhibitors of aggregation of isolated platelets in vitro.[135] However, following stimulation of platelet aggregation by ADP in vitro nitroglycerin has been shown to exert a potent disaggregating effect which may be due to a temporary elevation in the sensitivity of GC-S to NO.[136] Nitrate-induced inhibition of platelet aggregation in vitro can be greatly potentiated in the presence of thiols or cultured vascular cells.[134,137] This indicates that the conversion of organic nitrates by the vascular tissue in vivo can result in the release of sufficient amounts of NO for inhibition of platelet function. Indeed, in healthy volunteers, oral, transdermal and intravenous administrations of glyceryl trinitrate (nitroglycerin) and isosorbide mononitrates resulted in inhibition of platelet aggregation ex vivo.[138-142] In addition, glyceryl trinitrate and isosorbide dinitrate inhibited experimental thrombosis and reocclusion after thrombolysis in dogs and rats.[143,144] The effectiveness of organic nitrates as antithrombotics increases with the extent of vascular injury. In normal pigs, the deposition of platelets on arterial segments following injury using balloon angioplasty is inhibited by intravenous infusion of nitroglycerin when arterial injury is deep (extending through the internal elastic lamina) rather than mild (deendothelialization only).[145] Furthermore, short- and long-lasting administration of nitroglycerin and isosorbide dinitrate to patients suffering from coronary artery disease and acute myocardial infarction resulted in a significant inhibition of platelet adhesion and aggregation.[146-148] Of particular interest are interactions of organic nitrates with other inhibitors of platelet function. Organic nitrates synergize with prostacyclin or aspirin as inhibitors of platelet aggregation.[142,149] Also, isosorbide dinitrate potentiated the antiplatelet activity of prostaglandin E_1 in patients with peripheral vascular disease.[150]

The second group of NO donors is composed of compounds which release NO without the need of metabolic activation. Sodium nitroprusside, molsidomine and SIN-1 are clinically used members of this group.

Because of its powerful vasodilator action sodium nitroprusside is often used to treat vascular emergencies associated with hypertensive crisis. Since this compound shows some antiplatelet activity both in vitro and in vivo[151,152] its acute clinical effects may also be mediated, in part, through inhibition of platelet function.

Molsidomine and its active metabolite SIN-1 inhibit experimental thrombosis and platelet aggregation in healthy volunteers and in patients suffering from acute myocardial infarction.[153] Interestingly, SIN-1 in addition to NO generates superoxide and $ONOO^-$.[154] Since $ONOO^-$ causes platelet aggregation and counteracts the platelet inhibitory activity of NO,[52] the formation of this radical may offset the antiplatelet activity of NO released from SIN-1.

Despite almost 140 years of use, organic nitrates remain one of the safest and most widely applied cardiovascular drugs. However, since organic nitrates under some conditions are known to induce tachyphylaxis, there have been many attempts to synthesize tachyphylaxis-free NO donors. One of the more promising groups of drugs are cysteine-containing nitrates. The incorporation of a cellular thiol cysteine to the structure of organic nitrate resulted in a high effectiveness of these compounds as inhibitors of platelet and leukocyte functions both in vitro and in vivo.[155]

Another challenging problem is tissue-selectivity. The platelet-inhibitory actions of organic nitrates cannot be separated from their effects on the vascular wall. The concept of platelet-selective NO donors has arisen from our experiments with S-nitrosoglutathione (GSNO). S-nitrosoglutathione is a tripeptide S-nitrosothiol that is formed by S-nitrosylation of glutathione, the most abundant intracellular thiol. We have found that the intravenous administration of GSNO into conscious rats inhibits platelet aggregation at doses that have only a small effect on the blood pressure.[66] Moreover, similar platelet/vascular differentiation is detected following intraarterial administration of GSNO into the circulation of human forearm.[156] Finally, we have infused GSNO into patients undergoing balloon angioplasty and found that this NO donor effectively protected platelets from activation at the site of angioplastic injury without altering blood pressure.[157] Interestingly, the exposure of human neutrophils to NO led to depletion of glutathione stores, activation of hexose monophosphate shunt, synthesis of endogenous GSNO and inhibition of superoxide generation by neutrophils. Synthetic GSNO resulted in similar effects.[158] These observations show that GSNO is a potent regulator of platelet and neutrophil functions, and it may be a prototype for the development of blood cell-selective NO donors.

What is the position of organic nitrates among "classical" inhibitors of platelet function? Acteylsalicylic acid (aspirin) is by far the most widely used antiplatelet drug in clinical practice, and its benefits in terms of decreasing mortality due to reinfarction have been unequivocally demonstrated[159-162] while those of organic nitrates have not yet been established. A meta-analysis found significant reduction in mortality when intravenous glyceryl trinitrate or nitroprusside were used during acute course of myocardial infarction.[163] Moreover, when combined with N-acetylcysteine, glyceryl trinitrate substantially reduced myocardial infarction in unstable angina, an effect compatible with an antiplatelet effect of glyceryl trinitrate.[164] Surprisingly, GISSI III and ISIS-4 studies failed to show a clinically beneficial effect of organic nitrates on mortality after myocardial infarction.[165,166] However, further analysis of GISSI III suggests that the apparent additive effect of glyceryl trinitrate and lisinopril could be attributed to antiplatelet effects of this NO donor.[141] In addition, it is possible that nitrates may act by reducing the infarct size in small rather than large infarcts so that the neutral results of GISSI-3 and ISIS-4 may be explained by the heterogeneity of effect.[167] Interestingly aspirin, a cyclooxygenase inhibitor, blocks only thromboxane-mediated platelet aggregation[161] leaving the remaining pathways of adhesion and aggregation unopposed. In contrast, NO inhibits the activation cascade of mediators generated by all known pathways of platelet aggregation[168] and some pathways of platelet adhesion to subendothelium.[169] Inhibition of platelet adhesion may be of particular importance to decrease platelet accumulation due to ischemia-reperfusion injury following organ ischemia, and drugs such as NO donors that increase cGMP levels may afford effective antiplatelet action.[170] Thus, there is a clear need for further clinical studies to determine the place and the effectiveness of traditional organic nitrates and new NO donors for the treatment of vascular thrombotic and ischemic disorders.

CONCLUSIONS

Nitric oxide generated by endothelial cells, platelets and leukocytes has proved to be an important physiologic and regulatory mediator modulating vessel wall hemostasis and preventing thrombosis. The changes in its generation or metabolism are likely to play roles in the pathomechanism of vasospastic and thrombotic disorders. Understanding the regulatory and damaging properties of NO and its interaction with other hemostatic factors may help in improving current antithrombotic therapy.

ACKNOWLEDGEMENTS

This work was supported by establishment award from the Alberta Heritage Foundation for Medical Research. M. Radomski is a scholar of the Alberta Heritage Foundation for Medical Research.

REFERENCES

1. Slack MS, Cui Y, Turitto VT. The effects of flow on blood coagulation and thrombosis. Thromb Haemost 1993; 70:129-134.
2. Ginsberg MH, Loftus JC, Plow EF. Cytoadhesins, integrins and

platelets. Thromb Haemost 1988; 59:1-6.

3. Radomski MW. Platelet regulation-another dimension for NO and nitrates. Schwarz Pharma Scientific Forum, Schwarz Pharma Publisher 1994; vol 7.

4. Moncada S, Gryglewski RJ, Bunting S, Vane JR. An enzyme isolated from arteries transforms prostaglandin endoperoxides to an unstable substance that inhibits platelet aggregation. Nature 1976; 263:663-665.

5. Radomski MW, Palmer RMJ, Moncada S. Endogenous nitric oxide inhibits human platelet adhesion to vascular endothelium. Lancet 1987; 2:1057-1058.

6. Moncada S. Biological importance of prostacyclin. Br J Pharmacol 1982; 76:3-31.

7. Buchanan MR, Brister SJ. Antithrombotics and the lipoxygenase pathway. In: Herman AG, ed. Antithrombotics. Kluwer Academic Publishers, 1991; 159-180.

8. Marcus AJ, Safier LB. Thromboregulation: multicellular modulation of platelet reactivity in hemostasis and thrombosis. FASEB J 1993; 7:516-522.

9. Furchgott RF, Zawadzki JV. The obligatory role of endothelial cells in the relaxation of arterial smooth muscle by acetylcholine. Nature 1980; 288:373-376.

10. Furchgott RF. Studies on relaxation of rabbit aorta by sodium nitrite: The basis for the proposal that the acid-activatable inhibitory factor from bovine retractor penis is inorganic nitrite and that endothelium-derived relaxing factor is nitric oxide. In: Vanhoutte PM, ed. Vascular Smooth Muscle, Peptides, Autonomic Nerves, and Endothelium. Raven Press 1988; 401-414.

11. Ignarro LJ, Buga GM, Wood KS, Byrns RE, Chaudhuri G. Endothelium-derived relaxing factor produced and released from artery and vein is nitric oxide. Proc Natl Acad Sci USA 1987; 84:9625-9629.

12. Palmer RMJ, Ferrige AG, Moncada S. Nitric oxide release accounts for the biological activity of endothelium-derived relaxing factor. Nature 1987; 327:524-526.

13. Palmer RMJ, Ashton DS, Moncada S. Vascular endothelial cells synthesize nitric oxide from L-arginine. Nature 1988; 333:664-666.

14. Radomski MW. Nitric oxide-biological mediator, modulator and effector molecule. Ann Med 1995; in press.

15. Knowles RG. Nitric oxide synthases. The Biochemist 1994; 16:3-7.

16. Sessa WC. The nitric oxide synthase family of proteins. J Vasc Res 1994; 31:131-143.

17. Xie Q, Nathan C. The high-output nitric oxide pathway: role and regulation. J Leukoc Biol 1994; 56:576-582.

18. Radomski MW, Palmer RMJ, Moncada S. An L-arginine/nitric oxide pathway present in human platelets regulates aggregation. Proc Nat Acad Sci USA 1990; 87:5193-5197.

19. Radomski MW, Palmer RMJ, Moncada S. Characterization of the L-arginine: nitric oxide pathway in human platelets. Br J Pharmacol

1990; 101:325-328.

20. Goldberg ND, Haddox MK, Nicol SE et al. Biological regulation through opposing influences of cyclic GMP and cyclic AMP: The yin yang hypothesis. In: Drummond GI, Greengard P, Robinson GA eds. Advances in Cyclic Nucleotide Research, 1975; 5:307-330.

21. Mellion BT, Ignarro LJ, Ohlstein EH et al. Evidence for the inhibitory role of guanosine 3'-5'-monophosphate in ADP-induced human platelet aggregation in the presence of nitric oxide and related nitrovasodilators. Blood 1981; 57:946-955.

22. Caren R, Corbo L. Response of plasma lipids and platelet aggregation to intravenous arginine. Proc Soc Exp Biol Med 1973; 143:1067-1071.

23. Houston DS, Gerrard JM, McCrea J, Glover S, Butler AM. The influence of amines on various platelet responses. Biochim Biophys Acta 1983; 734:267-273.

24. Pronai L, Ichimori K, Nozaki H et al. Investigation of the existence and biological role of L-arginine/nitric oxide pathway in human platelets by spin-trapping/EPR studies. Eur J Biochem 1991; 202:923-930.

25. Noris M, Benigni A, Boccardo P et al. Enhanced nitric oxide synthesis in uremia: implications for platelet dysfunction and dialysis hypotension. Kidney Int 1993; 44:445-450.

26. Cadwgan TM, Benjamin N. Evidence for altered platelet nitric oxide synthesis in essential hypertension. J Hypertens 1993; 11:417-420.

27. Bodzenta-Lukaszyk A, Gabryelewicz A, Lukaszyk A et al. Nitric oxide synthase inhibition and platelet function. Thromb Res 1994; 75:667-672.

28. Muruganandam A, Mutus B. Isolation of nitric oxide synthase from human platelets. Biochim Biophys Acta 1994; 1200:1-6.

29. Nakatsuka M, Osawa Y. Selective inhibition of the 12-lipoxygenase pathway of arachidonic acid metabolism by L-arginine or sodium nitroprusside in intact human platelets. Biochem Biophys Res Commun 1994; 200:1630-1634.

30. Launay JM, Bondoux D, Oset-Gasque MJ et al. Increase of human platelet serotonin uptake by atypical histamine receptors. Am J Physiol 1994; 266:R526-R536.

31. Chen LY, Mehta JL. Inhibitory effect of high-density lipoprotein on platelet function is mediated by increase in nitric oxide synthase activity in platelets. Life Sci 1994; 23:1815-1821.

32. Zhou Q, Hellermann GR, Solomonson LP. Nitric oxide release from resting platelets. Thromb Res 1995; 77:87-96.

33. Mehta JL, Chen LY. Identification of constitutive and inducible forms of nitric oxide synthase in human platelets. J Lab Clin Med 1995; 125:370-377.

34. Golino P, Capelli-Bigazzi M, Ambrosio G et al. Endothelium-derived relaxing factor modulates platelet aggregation in an in vivo model of reccurent platelet activation. Circ Res 1992; 71:1447-1456.

35. Berkels R, Klaus W, Boller M, Rosen R. The calcium modulator nifedipine exerts its antiaggregatory property via a nitric oxide mediated process. Thromb Haemostas 1994; 72:309-312.

36. Yao SK, Ober JC, Krishnaswami A et al. Endogenous nitric oxide protects against platelet aggregation and cyclic flow variations in stenosed and endothelium-injured arteries. Circulation 1992; 86:1302-1309.

37. Thom SR, Ohnishi T, Ischiropoulos H. Nitric oxide released by platelets inhibits neutrophil B_2 integrin function following acute carbon monoxide poisoning. Toxicol Pharmacol 1994; 128:105-110.

38. Lelchuk R, Carrier M, Hancock V, Martin JF. The relationship between megakaryocyte nuclear DNA content and gene expression. Int J Cell Clon 1990; 8:277-282.

39. Lelchuk R, Radomski MW, Martin JF, Moncada S. Constitutive and inducible nitric oxide synthases in human megakaryoblastic cells. J Pharmacol Exp Ther 1992; 262:1220-1224.

40. Djaffar I, Vilette D, Bray PF, Rosa JP. Quantitative isolation of RNA from human platelets. Thromb Res 1991; 62:127-135.

41. Durante W, Schini VB, Kroll MH et al. Platelets inhibit the induction of nitric oxide synthesis by interleukin-1β in vascular smooth muscle cells. Blood 1994; 83:1831-1838.

42. Hawrylowicz CM, Howells GL, Felmann M. Platelet-derived interleukin induces human endothelial adhesion molecule expression and cytokine production. J Exp Med 1991; 174:785-790.

43. Marsden PA, Shappert KT, Chen HS et al. Molecular cloning and characterization of human endothelial nitric oxide synthase. FEBS Lett 1992; 307:287-293.

44. Sessa WV, Harrison JK, Barber CM et al. Molecular cloning and expression of cDNA encoding endothelial cell nitric oxide synthase. J Biol Chem 1992; 267:15274-15276.

45. Radomski MW, Palmer RMJ, Moncada S. Glucocorticoids inhibit the expression of an inducible but not the constitutive, nitric oxide synthase in vascular endothelial cells. Proc Natl Acad Sci USA 1990; 87:10043-10047.

46. Kanno K, Hirata Y, Taihei I, Iwashina M, Marumo F. Regulation of inducible nitric oxide synthase gene by interleukin-1β in rat vascular endothelial cells. Am J Physiol 1994; 267:H2318-H2324.

47. Gross SS, Jaffe E, Levi R, Kilbourn RG. Cytokine-activated endothelial cells express an isotype of nitric oxide synthase which is tetrahydrobiopterin-dependent, calmodulin-independent and inhibited by arginine analogs with a rank-order of potency characteristic of activated macrophages. Biochem Biophys Res Comun 1991; 178:823-829.

48. Vanderkooi JM, Wright WW, Erecinska M. Nitric oxide diffusion coefficients in solutions, proteins and membranes determined by

phosphorescence. Biochim Biophys Acta 1994; 1207:249-254.

49. Taha Z, Kiechle F, Malinski T. Oxidation of nitric oxide by oxygen in biological systems monitored by porphyrinic microsensor. Biochem Biophys Res Commun 1992; 188:734-739.

50. Beckman J, Tsai JH. Reactions and diffusion of nitric oxide and peroxynitrite. The Biochemist 1994; 16:8-10.

51. Lipton SA, Choi YB, Pan ZH et al. A redox-based mechanism for the neuroprotective and neurodestructive effects of nitric oxide and related nitroso-compounds. Nature 1993; 364:626-632.

52. Moro MA, Darley-Usmar VM, Goodwin DA et al. Paradoxical fate and biological action of peroxynitrite on human platelets. Proc Natl Acad Sci USA 1994; 91:6702-6706.

53. Murphy ME, Sies H. Reversible conversion of nitroxyl anion to nitric oxide by superoxide dismutase. Proc Natl Acad Sci USA 1991; 88:10860-10864.

54. Fukuto JM, Chiang K, Hszieh R, Wong P, Chaudhuri G. The pharmacological activity of nitroxyl: a potent vasodilator with activity similar to nitric oxide and/or endothelium-derived relaxing factor. J Pharmacol Exp Ther 1992; 263:546-551.

55. Craven PA, DeRubertis FR. Restoration of the reponsiveness of purified guanylate cyclase to nitrosoguanidine, nitric oxide, and related activators by heme and heme proteins. Evidence for involvement of the paramagnetic nitrosyl-heme complex in enzyme activation. J Biol Chem 1978; 253:8433-8443.

56. Liu Z, Nakatsu K, Brien JF et al. Selective sequestration of nitric oxide by subcellular components of vascular smooth muscle and platelets: relationship to nitric oxide stimulation of the soluble guanylyl cyclase. Can J Physiol Pharmacol 1993; 71:938-945.

57. Guthmann F, Mayer B, Koesling D, Kukovetz WR, Bîhme E. Characterization of soluble guanylyl cyclase with peptide antibodies. Naunyn-Schmiedeberg's Arch Pharmacol 1992; 346:537-541.

58. Buechler WA, Ivanova K, Wolfram G et al. Soluble guanylyl cyclase and platelet function. In: FitzGerald GA, Jennings LK, Patrono C eds. Annals of the New York Academy of Sciences 1994; 714:151-157.

59. Walter U. Physiological role of cGMP and cGMP-dependent protein kinase in the cardiovascular system. Rev Physiol Biochem Pharmacol 1989; 113:41-88.

60. HalbrÅgge M, Walter U. Purification of a vasodilator-regulated phosphoprotein from human platelets. Eur J Biochem 1989; 185:41-50.

61. Reinhard M, HalbrÅgge M, Scheer U et al. The 46/50 kDa phosphoprotein VASP purified from human platelets is a novel protein associated with actin filaments and focal contacts. EMBO J 1992; 11:2063-2070.

62. Horstrup K, Jablonka B, Hînig-Liedl P et al. Phosphorylation of focal adhesion vasodilator-stimulated phosphoprotein at Ser157 in intact human platelets correlates with fibrinogen receptor inhibition. Eur J Biochem 1994; 225:21-27.

63. Nakashima S, Tohmatsu T, Hattori H, Okano Y, Nozawa Y. Inhibitory action of cyclic GMP on secretion, phosphoinositide hydrolysis and calcium mobilization in thrombin-stimulated human platelets. Biochem Biophys Res Commun 1986; 135:1099-1104.

64. Johansson JS, Hayness DH. Cyclic GMP increases the rate of the calcium extrusion pump in intact platelets but has no direct effect on the dense tubular calcium accumulation system. Biochim Biophys Acta 1992; 1105:40-50.

65. Maurice DH, Haslam RJ. Molecular basis of the synergistic inhibition of platelet function by nitrovasodilators and activators of adenylate cyclase: inhibition of cyclic AMP breakdown by cyclic GMP. Mol Pharmacol 1990; 37:671-681.

66. Radomski MW, Rees DD, Dutra A, Moncada S. S-Nitrosoglutathione inhibits platelet activation in vitro and in vivo. Br J Pharmacol 1992; 107:745-749.

67. Sane DC, Bielawska A, Greenberg CS, Hannun YA. Cyclic GMP analogs inhibit gamma thrombin-induced arachidonic acid release in human platelets. Biochem Biophys Res Commun 1989; 165:708-714.

68. Salas E, Moro MA, Askew S et al. Comparative pharmacology of analogues of S-nitroso-N-acetyl-DL-penicillamine in platelets. Br J Pharmacol 1994; 112:1071-1076.

69. Mendelsohn ME, O'Neill S, George D, Loscalzo J. Inhibition of fibrinogen binding to human platelets by S-nitroso-N-acetylcysteine. J Biol Chem 1990; 265:19028-19034.

70. Wu XB, Brune B, von Appen F, Ullrich V. Efflux of cyclic GMP from activated platelets. Mol Pharmacol 1993; 43:564-568.

71. Brune B, Lapetina EG. Activation of a cytosolic ADP-ribosyltransferase by nitric oxide-generating agents. J Biol Chem 1989; 264:8455-8458.

72. McDonald B, Reep B, Lapetina EG, Molina y Vedia L. Glyceraldehyde-3-phosphate dehydrogenase is required for the transport of nitric oxide in platelets. Proc Natl Acad Sci USA 1993; 90:11122-11126.

73. Ambrosio G, Golino P, Pascucci I et al. Modulation of platelet function by reactive oxygen metabolites. Am J Physiol 1994; 267:H308-H318.

74. Brune B, Ullrich V. 12-Hydroperoxyeicosatatraenoic acid inhibits main platelet function by activation of soluble guanylate cyclase. Mol Pharmacol 1991; 39:671-678.

75. Brune B, Ullrich V. Inhibition of platelet aggregation by carbon monoxide is mediated by activation of guanylate cyclase. Mol Pharmacol 1987; 32:497-504.

76. Stamler JS, Simon DJ, Osborne JA et al. S-Nitrosylation of proteins with nitric oxide: synthesis and characterization of biologically active compounds. Proc Natl Acad Sci USA 1992; 89:444-448.

77. Radomski MW, Palmer RMJ, Moncada S. Comparative pharmacology of endothelium-derived relaxing factor, nitric oxide and

prostacyclin in platelets. Br J Pharmacol 1987; 92:181-187.

78. Malinski T, Radomski MW, Taha Z, Moncada S. Direct electrochemical measurement of nitric oxide released from human platelets. Biochem Biophys Res Commun 1993; 194:960-965.

79. Polanowska-Grabowska R, Gear ARL. Role of cyclic nucleotides in rapid platelet adhesion to collagen. Blood 1994; 83:2508-2515.

80. Brune B, Volker U. Different calcium pools in human platelets and their role in thromboxane A_2 formation. J Biol Chem 1991; 266:19232-19237.

81. Rees DD, Palmer RMJ, Moncada S. Role of endothelium-derived nitric oxide in the regulation of blood pressure. Proc Natl Acad Sci USA 1989; 86:3375-3378.

82. Cooke JP, Rossitch Jr E, Andon NA, Loscalzo J, Dzau VJ. Flow activates an endothelial potassium channel to release an endogenous nitrovasodilator. J Clin Invest 1991; 88:1663-1671.

83. Radomski MW, Palmer RMJ, Moncada S. The role of nitric oxide and cGMP in platelet adhesion to vascular endothelium. Biochem Biophys Res Commun 1987; 148:1482-1489.

84. Sneddon JM, Vane JR. Endothelium-derived relaxing factor reduces platelet adhesion to bovine endothelial cells. Proc Natl Acad Sci USA 1988; 85:2800-2804.

85. Venturini CM, Del Vecchio PJ, Kaplan JE. Thrombin-induced platelet adhesion to endothelium is modified by endothelial derived relaxing factor (EDRF). Biochem Biophys Res Commun 1989; 159:349-354.

86. Pohl U, Busse R. EDRF increases cyclic GMP in platelets during passage through the coronary vascular bed. Circ Res 1989; 65:1798-1803.

87. Radomski MW, Palmer RMJ, Moncada S. The anti-aggregating properties of vascular endothelium: interactions between prostacyclin and nitric oxide. Br J Pharmacol 1987; 92:639-646.

88. Furlong B, Henderson AH, Lewis MJ, Smith JA. Endothelium-derived relaxing factor inhibits in vitro platelet aggregation. Br J Pharmacol 1987; 90:687-692.

89. Busse R, Luckhoff A, Bassenge E. Endothelium-derived relaxing factor inhibits platelet activation. Naunyn Schmiedeberg's Arch Pharmacol 1987; 336:566-571.

90. Macdonald PS, Read MA, Dusting GJ. Synergistic inhibition of platelet aggregation by endothelium-derived relaxing factor and prostacyclin. Thromb Res 1988; 49:437-449.

91. Alheid U, Reichwehr I, Forstermann U. Human endothelial cells inhibit platelet aggregation by separately stimulating platelet cyclic. AMP and cyclic GMP. Eur J Pharmacol 1989; 164:103-110.

92. Houston DS, Robinson P, Gerrard JM. Inhibition of intravascular platelet aggregation by endothelium-derived relaxing factor: reversal by red blood cells. Blood 1990; 76:953-958.

93. Broekman MJ, Eiroa AM, Marcus AJ. Inhibition of human plate-

let reactivity by endothelium-derived relaxing factor from human umbilical vein endothelial cells in suspension: blockade of aggregation and secretion by an aspirin-insensitive mechanism. Blood 1991; 78:1033-1040.

94. May GR, Crook P, Moore PK, Page CP. The role of nitric oxide as an endogenous regulator of platelet and neutrophil activation within the pulmonary circulation. Br J Pharmacol 1991; 102:759-763.

95. Herbaczynska-Cedro K, Lembowicz K, Pytel B. N_G-monomethyl-L-arginine increases platelet deposition on damaged endothelium in vivo. A scanning electron microscopy study. Thromb Res 1991; 64:1-9.

96. Rosenblum WI, Nelson GH, Povlishock JT. Laser-induced endothelial damage inhibits endothelium-dependent relaxation in the cerebral microcirculation of the mouse. Circ Res 1987; 60:169-176.

97. Bhardwaj R, Page CP, May GR, Moore PK. Endothelium-derived relaxing factor inhibits platelet aggregation in human whole blood in vitro and in the rat in vivo. Eur J Pharmacol 1988; 157:83-91.

98. Hogan JC, Lewis MJ, Henderson AH. In vivo EDRF activity influences platelet function. Br J Pharmacol 1988; 94:1020-1022.

99. Humphries RG, Tomlinson W, O'Connor SE, Leff P. Inhibition of collagen- and ADP-induced platelet aggregation by substance P in vivo: Involvement of endothelium-derived relaxing factor. J Cardiovasc Pharmacol 1990;16:292-297.

100. Houston DS, Buchanan MR. Influence of endothelium-derived relaxing factor on platelet function and hemostasis in vivo. Thromb Res 1994; 74:25-37.

101. Andrews NP, Dakak N, Schenke WH, Quyyumi AA. Platelet-endothelium interactions in humans: changes in platelet cyclic guanosine monophosphate content in patients with endothelial dysfunction.Circulation 1994; 90:I-397.

102. Adams MR, Forsyth C, Robinson J, Jessup W, Celermajer DS. Oral L-arginine in humans: effects on platelet aggregation, hemodynamics and endothelium-dependent dilatation. Circulation 1994; 90:1-138.

103. McCall TB, Boughton-Smith NK, Palmer RMJ, Whittle BJR, Moncada S. Synthesis of nitric oxide from L-arginine by neutrophils. Biochem J 1989; 261:293-296.

104. Kubes P, Suzuki M, Granger DN. Nitric oxide: an endogenous modulator of leukocyte adhesion. Proc Natl Acad Sci USA 1991; 88:4651-4655.

105. Moilanen E, Vuorinen P, Metsa-Ketela T, Vapaatalo H. Inhibition by nitric oxide donors of human polymorphonuclear leucocyte functions. Br J Pharmacol 1993; 109:852-858.

106. Stamler JS, Vaughan DE, Loscalzo J. Synergistic disaggregation of platelets by tissue-type plasminogen activator, prostaglandin E_1 and glyceryl trinitrate. Circ Res 1989; 65:796-804.

107. de Belder AJ, Radomski MW. Nitric oxide in the clinical arena. J Hypertens 1994; 12:617-624.

108. Luscher TF, Tanner FC, Tschudi MR, Noll G. Endothelial dysfunction in coronary artery disease. Ann Rev Med 1993; 44:395-418.
109. Flavahan NA. Atherosclerosis or lipoprotein-induced endothelial dysfunction.Circulation 1992; 85:1927-1938.
110. Cooke JP, Tsao P. Cellular mechanisms of atherogenesis and the effects of nitric oxide. Curr Opin Cardiol 1992; 7:799-804.
111. Chen L, Mehta JL. High density lipoprotein antagonizes the stimulatory effect of low density lipoprotein on platelet function by affecting L-arginine-nitric oxide pathway. Circulation 1994; 90 (4):I-30.
112. Tsao PS, Theilmeier G, Singer AH, Leung LLK, Cooke JP. L-Arginine attenuates platelet reactivity in hypercholesterolemic rabbits. Arterioscler Thromb 1994; 14:1529-1533.
113. Calver A, Collier J, Moncada S, Vallance P. Effect of local infusion of N_G-monomethyl-L-arginine in patients with hypertension. The nitric oxide dilator mechanism appears abnormal. J Hypertens 1992; 10:1025-1031.
114. Amado JA, Salas E, Botana MA, Poveda JJ, Berrazueta JR. Low levels of intraplatelet cGMP in IDDM. Diabetes Care 1993; 16: 809-811.
115. Komamura K, Node K, Kosaka H, Inoue M. Endogenous nitric oxide inhibits microthromboembolism in the ischemic heart. Circulation 1994; 90:I-345.
116. Olsen SB, Ayala B, Tang DB et al. Enhancement of platelet deposition by cross-linked hemoglobin in a rat carotid endarterectomy model. Circulation 1994; 90:I-345.
117. Myers PR, Wright TF, Tanner MA, Ostlund Jr RE. The effects of native LDL and oxidized LDL on EDRF bioactivity and nitric oxide production in vascular endothelium. J Lab Clin Med 1994; 124:672-683.
118. Kooy NW, Royall JA. Agonist-induced peroxynitrite production from endothelial cells. Arch Biochem Biophys 1994; 310:352-359.
119. Villa LM, Salas E, Darley-Usmar VM, Radomski MW, Moncada S. Peroxynitrite induces both vasodilatation and impaired vascular relaxation in the isolated perfused rat heart. Proc Natl Acad Sci USA 1994; 91:12383-12387.
120. White CR, Brock TA, Chang LY et al. Superoxide and peroxynitrite in atherosclerosis. Proc Natl Acad Sci USA 1994; 91:1044-1048.
121. Radomski MW, Vallance P, Whitley G, Foxwell N, Moncada S. Platelet adhesion to human vascular endothelium is modulated by constitutive and cytokine induced nitric oxide. Cardiovasc Res 1993; 27:1380-1382.
122. Schultz PJ, Raij L. Endogenously synthesized nitric oxide prevents-endotoxin-induced glomerular thrombosis. J Clin Invest 1992; 90:1718-1725.
123. Spain DA, Wilson MA, Garrison RN. Nitric oxide synthase inhibition exacerbates sepsis-induced renal hypoperfusion. Surgery 1994;

116:322-331.

124. Palmer RMJ, Bridge L, Foxwell NA, Moncada S. The role of nitric oxide in endothelial cell damage and its inhibition by glucocorticoids. Br J Pharmacol 1992; 105:11-12.

125. Kilbourn RG, Gross SS, Adams J et al. N_G-methyl-L-arginine inhibits tumor necrosis factor-induced hypotension: implications for the involvement of nitric oxide. Proc Natl Acad Sci USA 1990; 87:3029-3032.

126. Wright CE, Rees DD, Moncada S. Protective and pathological role of nitric oxide in endotoxin shock. Cardiovasc Res 1992; 26:48-57.

127. Horowitz HI, Stein IM, Cohen BD, White JG. Further studies on the platelet-inhibitory effect of guanidinosuccinic acid and its role in uremic bleeding. Am J Med 1970; 49:336-345.

128. Radomski MW, Jenkins DC, Holmes L, Moncada S. Human colorectal adenocarcinoma cells: differential nitric oxide synthesis determines their ability to aggregate platelets. Cancer Res 1991; 51:6073-6078.

129. Dong Z, Staroselsky AH, Qi X, Xie K, Fiedler I. Inverse correlation between expression of inducible nitric oxide synthase activity and production of metastasis in K-1735 murine melanoma cells. Cancer Res 1994; 54:789-793.

130. Thomsen L, Lawton FG, Knowles RG et al. Nitric oxide synthase activity in human gynecological cancer. Cancer Res 1994; 54:1352-1354.

131. Hogman M, Frostell C, Arnberg H, Hedenstierna G. Bleeding time prolongation and NO inhalation. Lancet 1993; 341:1664-1665.

132. Malinski T, Taha Z, Grunfeld S et al. Diffusion of nitric oxide in the aorta wall monitored in situ by porphyrinic microsensors. Biochem Biophys Res Commun 1993; 193:1076-1082.

133. Feelisch M, Noack EA. Correlation between nitric oxide formation during degradation of organic nitrates and activation of guanylyl cyclase. Eur J Pharmacol 1987; 139:19-30.

134. Feelisch M. The action and metabolism of organic nitrates and their similarity with endothelium-derived relaxing factor (EDRF). In: Moncada S, Higgs EA, Berrazueta JR eds. Clinical Relevance of Nitric Oxide in the Cardiovascular System, Edicomplet Madrid 1991; 29-43.

135. Gerzer R, Karrenbrock B, Siess W, Heim JM. Direct comparison of the effects of nitroprusside, SIN-1 and various nitrates on platelet aggregation and soluble guanylyl cyclase activity. Thromb Res 1988; 52:11-21.

136. Chirkov YY, Naujalis R, Sage E, Horowitz YD. Antiplatelet effects of nitroglycerin in healthy subjects and in patients with stable angina pectoris. J Cardiovasc Pharmacol 1993; 21:384-383.

137. Benjamin N, Dutton JAE, Ritter JM. Human vascular smooth muscle cells inhibit platelet aggregation when incubated with glyceryl trinitrate: evidence for generation of nitric oxide. Br J Pharmacol 1991; 102:847-850.

138. de Caterina T, Giannesssi D, Crea F et al. Inhibition of platelet function by injectable isosorbide dinitrate. Am J Cardiol 1984; 53:1683-1687.
139. Drummer C, Valta-Seufzer U, Karrenbrock B, Heim JM, Gerzer R. Comparison of antiplatelet properties of molsidomine, isosorbide-5-mononitrate and placebo in healthy volunteers. Eur Heart J 1991; 12:541-549.
140. Wallen NH, Larsson PT, Broijersen A, Andersson A, Hjemdahl P. Effects of oral dose of isosorbide dinitrate on platelet function and fibrinolysis in healthy volunteers. Br J Clin Pharmac 1994; 72:575-579.
141. Andrews R, May YA, Vickers L, Heptinstall S. Inhibition of platelet aggregation by transdermal glyceryl trinitrate. Br Heart J Cardiovasc Pharmacol 1993; 21:384-389.
142. Karlberg KE, Ahlner J, Henriksson P, Torfgard K, Sylven C Effects of nitroglycerin on platelet aggregation beyond the effects of acetylsalicylic acid in healthy subjects. Am J Cardiol 1993; 71:361-364.
143. Werns SW, Rote WE, Davis JH, Guevara T, Lucchesi BR. Nitroglycerin inhibits experimental thrombosis and reocclusion after thrombolysis. Am Heart J 1994; 127:727-737.
144. Plotkine M. Allix M, Guillou J, Boulu R. Oral administration of isosorbide dinitrate inhibits arterial thrombosis in rats. Eur J Pharmacol 1991; 201:115-116.
145. Lam JYT, Chesebro JH, Fuster V. Platelets, vasoconstriction, and nitroglycerin during arterial wall injury. A new antithrombotic role for an old drug. Circulation 1988; 78:712-716.
146. Gebalska J. Platelet adhesion and aggregation in relation to clinical course of acute myocardial infarction. M.D. thesis 1990; Warsaw, in Polish.
147. Diodati J, Theroux P, Latour JG et al. Effects of nitroglycerin at therapeutic doses on platelet aggregation in unstable angina pectoris and acute myocardial infarction. Am J Cardiol 1990; 66:683-688.
148. Sinzinger H, Virgolini I, O'Grady J, Rauscha F, Fitscha P. Modification of platelet function by isosorbide dinitrate in patients with coronary artery disease. Thromb Res 1992; 65:323-335.
149. de Caterina R, Giannessi, Bernini W, Mazzone A. Organic nitrates: direct antiplatelet effects and synergism with prostacyclin. Antiplatelet effects of organic nitrates. Thromb Haemost 1988; 59:207-211.
150. Sinzinger H, Fitscha P, O'Grady J et al. Synergistic effect of prostaglandin E1 and isosorbide dinitrate in peripheral vascular disease. Lancet 1990; 335:627-628.
151. Levin RL, Weksler BB, Jaffe EA. The interaction of sodium nitriprusside with human endothelial cells and platelets: nitroprusside and prostacyclin synergistically inhibit platelet function. Circulation 1982; 66:1299-1307.
152. Hines R, Barash PG. Infusion of sodium nitroprusside induces platelet dysfunction in vitro. Anesthesiology 1989; 71:805-806.

153. Wautier JL, Weill D, Kadeva H, Maclouf J, Soria C. Modulation of platelet function by SIN-1A. J Cardiovasc Pharmacol 1989; 14:S111-S114.

154. Hogg N, Darley-Usmar VM, Wilson MT, Moncada S. Oxidation of alpha-tocopherol in human low density lipoprotein by the simultaneous generation of superoxide and nitric oxide. FEBS Lett 1993; 326:199-203.

155. Lefer DJ, Nakanishi K, Vinten-Johansen J. Endothelial and myocardial cell protection by a cysteine-containing nitric oxide donor after myocardial ischemia and reperfusion. J Cardiovasc Pharmacol 1993; 22(Suppl. 7):S34-S43.

156. de Belder AJ, MacAllister R, Radomski MW, Moncada S, Vallance PJ. Effects of S-nitrosoglutathione in the human forearm circulation. Evidence for selective inhibition of platelet activation. Cardiovasc Res 1994; 28:691-694.

157. Langford EJ, Brown AS, Wainwright RJ et al. Inhibition of platelet activity by S-nitrosoglutathione during coronary angioplasty. Lancet 1994; 344:1458-1460.

158. Clancy RM, Levartovsky D, Leszczynska-Piziak J, Yegudin J, Abramson SB. Nitric oxide reacts with intracellular glutathione and activates the hexose monophosphate shunt in human neutrophils: evidence for S-nitrosoglutathione as a bioactive intermediary. Proc Natl Acad Sci USA 1994; 91:3680-3684.

159. ISIS-2 (Second International Study of Infarct Survival). Collaborative Group: Randomized trial of intravenous streptokinase, oral aspirin, both or neither among 17,187 cases of suspected acute myocardial infarction: ISIS-2. Lancet 1988; 2:349-360.

160. ISIS-3, a randomized comparison of streptokinase vs tissue plasminogen activator vs anistreplase and of aspirin plus heparin vs aspirin alone among 41,299 cases of suspected acute myocardial infarction. Lancet 1992; 339:753-770.

161. Patrono C. Aspirin and human platelets: from clinical trials to acetylation of cyclooxygenase and back. Trends Pharmacol Sci 1989; 10:453-458.

162. Fernandez-Ortiz A, Jang IK, Fuster A. Antiplatelet and antithrombin therapy. Coronary Artery Disease 1994; 5:297-305.

163. Yusuf S, MacMahon S, Collins R, Peto R. Effect of intravenous nitrates on mortality in acute myocardial infarction: an overview of the randomised trials. Lancet 1988; i:1088-1092.

164. Horowitz JD, Henry CA, Syrjanen ML et al. Combined use of nitroglycerin and N-acetylcysteine in the management of unstable angina pectoris. Circulation 1988; 77:787-794.

165. GISSI III study group. GISSI-3: effects of lisinopril and transdermal glyceryl trinitrate singly and together on 6-week mortality and ventricular function after acute myocardial infarction. Lancet 1994; 343:1115-1122.

166. ISIS collaborative group, Oxford, U.K. ISIS-4: randomised study of oral isosorbide mononitrate in over 50,000 patients with suspected acute myocardial infarction. Circulation 1993; 88:1-394.
167. Morris JL, Zaman AG, Smyllie JH, Cowan JC. The effect of intravenous nitrate on infarct size; evidence of no benefit in small but not large infarcts. Br Heart J 1994; 71:77.
168. Jensen BO, Holmsen H. Nitric oxide (NO)-platelet interactions: inhibition is independent of the prostanoid and ADP pathways. Abstracts of the Scandinavian Physiological Society, Bergen, Norway 1994; C2.
169. Shahbazi T, Jones N, Radomski MW, Moro MA, Gingell D. Nitric oxide donors inhibit platelet spreading on surfaces coated with fibrinogen but not fibronectin. Thromb Res 1994; 75:631-642.
170. Chintala MS, Bernardino V, Chiu PJ. Cyclic GMP but not cyclic AMP prevents renal platelet accumulation after ischemia-reperfusion in anesthetized rats. J Pharmacol Exp Ther 1994; 271:1203-1208.

MICROVASCULAR PERMEABILITY ALTERATIONS AND THE ROLE OF NITRIC OXIDE

Paul Kubes

INTRODUCTION

For many years, the endothelium lining the microvasculature was considered to be a passive filter with static permeability characteristics. Therefore, fluxes of fluid and protein out of the vasculature were considered to be primarily regulated by alterations in blood flow, microvascular pressure and the amount of perfused surface area and not an actual alteration in the size of pores between endothelial cells. Over the last 20 years, much evidence has accumulated to challenge the concept of a microvascular barrier with exclusively fixed porosity characteristics (reviewed in refs. 1-3). In fact, based on macromolecular transport studies under basal and mediator-induced states, it is now believed that endothelial permeability can be rapidly and reversibly induced to change. Majno et al[4] provided some of the earliest morphologic evidence that endothelial cells could be induced to separate thereby providing greater permeability to plasma macromolecules. Histamine induced active endothelial cell contracture in postcapillary venules that allowed for increased leakage of protein out of the vasculature. These data implied that the endothelium is a malleable barrier, and alterations in membrane porosity were a likely event during

Nitric Oxide: A Modulator of Cell-Cell Interactions in the Microcirculation, edited by Paul Kubes. © 1995 R.G. Landes Company.

inflammatory conditions. Prostaglandins released constitutively from endothelial cells have been purported to affect microvascular permeability. These agents including PGE_1 and PGE_2 are themselves weak modulators of microvascular permeability but can greatly potentiate microvascular permeability induced by other molecules.[5,6] There is some evidence that catecholamines may also modulate microvascular permeability. β_2-receptor agonists have been shown to reduce histamine-induced increase in microvascular permeability suggesting that catecholamines may also reduce microvascular permeability during inflammation.[7] Although much emphasis has been given to studying permeability alterations related to the inflammatory or mediator-stimulated pathway, the endogenous (moment-to-moment) physiologic regulation of microvascular permeability to macromolecules has remained largely ignored. Until recently, a pure endogenous modulator of microvascular permeability was merely an interesting but unsubstantiated concept.

Recently, it has been reported that endothelium-derived relaxing factor or nitric oxide (synthesized from a guanidino group of L-arginine) may play a very important role as an intrinsic modulator of microvascular permeability in the cat small intestine.[8] This conclusion was based on the fact that the analog of L-arginine, N^G-nitro-L-arginine methyl ester (L-NAME), which competes for the enzyme nitric oxide synthase and blocks nitric oxide production, caused a rapid increase in protein and fluid flux from the vasculature into the interstitium. Therefore, these data were consistent with the view that nitric oxide played a homeostatic regulatory role in fluid and protein movement across the vasculature. Discussing the mechanisms (including leukocyte adhesion) underlying the nitric oxide-induced microvascular permeability alterations are the aim of this chapter.

NITRIC OXIDE SYNTHESIS INHIBITION AFFECTS MICROVASCULAR PERMEABILITY

Inhibition of NO synthesis with the L-arginine analog L-NAME in the cat intestine[8] resulted in a very rapid and persistent increase in microvascular protein and fluid leakage out of the vasculature (Fig. 4.1). D-NAME the biologically inert enantiomer did not have any effect on the microvasculature suggesting that the increased transvascular fluid and protein flux associated with L-NAME was not related to a nonspecific (eg., cationic) effect of the nitric ox-

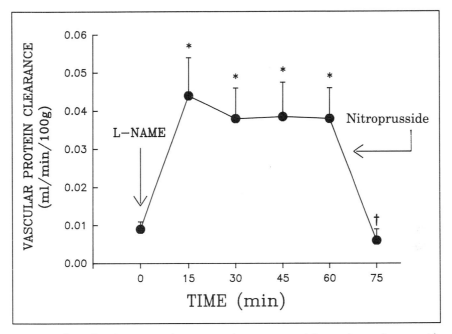

Fig. 4.1. The role of nitric oxide as an endogenous modulator of vascular protein clearance from the vasculature. L-NAME rapidly increased vascular protein leakage (measurement given after 60 mins) an event that was entirely reversible with the nitric oxide donor, nitroprusside. *p < 0.05 relative to time 0. † p < 0.05 relative to 60 mins value.

ide synthesis inhibitor. The L-NAME effect could be prevented by L-arginine but only if the dose of the nitric oxide precursor was given at five times the dose of L-NAME. Therefore sufficient amounts of L-arginine could overcome the effects of the NO synthesis inhibitor. Additionally, exogenous administration of nitric oxide via the NO donor nitroprusside also prevented the increased vascular protein and fluid efflux associated with prevention of NO synthesis inhibition (Fig. 4.1). These data suggest that nitroprusside was able to deliver sufficient amounts of NO to target cells to prevent the normally observed effects of L-NAME. Because of potential nonspecific effects of nitroprusside, in a subsequent study animals were pretreated with a second NO donor, SIN-1. L-NAME did not increase the leakage of protein and fluid out of the intestinal vasculature in the presence of this NO donor.[9] Interestingly, unlike the pretreatment regimen for L-arginine necessary to prevent the vascular effects of L-NAME, pretreatment with the NO donors was not necessary to reverse the rise in protein and fluid leakage induced by L-NAME. Local infusion of nitroprusside to the intestinal circulation following 60 mins of L-NAME exposure rapidly returned the elevated transvascular protein and fluid flux towards baseline values.[8] The reversal with NO donors occurred during continued administration of L-NAME to the intestinal circulation.

Despite the substantial elevation in protein and fluid efflux out of the microcirculation associated with NO synthesis inhibition, one could argue that this response was independent of microvascular permeability alterations. A simple increase in microvascular hydrostatic pressure could account for increased fluid and protein leakage out of the intestinal microcirculation associated with L-NAME. This was a concern because of the profound changes in hemodynamic alterations reported by many investigators following inhibition of NO synthesis. Therefore, intestinal capillary and venule pressures were measured directly in the presence and absence of L-NAME.[8] Capillary pressure decreased slightly from 9.1 ± 0.8 to 8.8 ± 0.5 mmHg at 60 mins of L-NAME administration. This is primarily as a result of a larger increase in precapillary vs postcapillary resistance. These measurements did not agree with the concept of an increase in microvascular hydrostatic pressure underlying the elevation in fluid and protein efflux out of the vasculature.

Another possibility was that L-NAME infusion led to the recruitment of capillaries thereby increasing surface area for protein and fluid filtration out of the microvasculature. However this possibility seemed untenable in light of the fact that vasoconstrictors such as L-NAME generally cause derecruitment of mesenteric capillaries. Nevertheless, we[8] directly tested the possibility that NO synthesis inhibition increases microvascular permeability not surface area by measuring osmotic reflection coefficient (σ_d). We eliminated the effect of changes in surface area and capillary pressures as confounding variables by achieving a filtration-independent steady-state condition. Venous pressure within the intestine was increased in 10 mmHg increments up to approximately 35 mmHg, an intervention that increased lymph flow (>10-fold) to a level where further changes in lymph flow no longer affected lymph to plasma protein concentration (filtration-independent). Under these filtration-independent conditions alterations in the movement of plasma protein assessed as lymph (C_L) to plasma (C_P) protein concentrations are entirely dependent upon pore size,[10] and $1-C_L/C_P$ provides an estimate of the osmotic reflection coefficient (σ_d). If the vasculature were entirely impermeable to plasma proteins σ_d

would approach 1, whereas if there was no barrier to the transvascular movement of plasma proteins, σ_d would be 0. Under control conditions in our intestinal preparation $1\text{-}\sigma_d$ (an index of permeability) was approximately 0.21 (Fig. 4.2). L-NAME doubled the restrictive properties of the vascular barrier (Fig. 4.2), a value previously reported for inflammatory conditions such as ischemia/reperfusion.[11] These data suggest a very profound increase in microvascular permeability following NO synthesis inhibition based on a measurement that excludes complications such as changes in microvascular pressures or surface area.

More recently, Prado et al[12] reported that inhibition of nitric oxide synthesis prompted the movement of macromolecules across the blood-brain barrier. These investigators found that infusion of L-NAME resulted in focal areas of extravasation of horseradish peroxidase within cortical regions. The increased leakage of HRP was observed in cortical arterioles, with HRP reaction product found within endothelial cell pinocytotic vesicles, smooth muscle basal laminae and the surrounding interstitial compartment. These data raised the interesting possibility that endogenous production

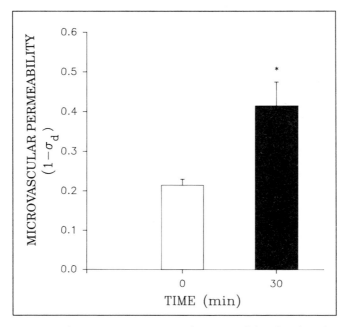

Fig. 4.2. Alterations in microvascular permeability (1-σ_d) under control conditions and following inhibition of nitric oxide synthase for 30 mins. *p < 0.05 relative to control (time 0).

of nitric oxide also contributes to the regulation of fluid and protein exchange across the cerebral microvasculature. Although increased vascular pressure could again be an underlying reason for the increased protein leakage in the cerebral vasculature, indirect observations did not support this hypothesis. It has been previously reported that agents that increase blood pressure (much like L-NAME) including norepinephrine and angiotensin as well as nonpharmacologically induced hypertension (abdominal aorta occlusion), also increased cerebrovascular permeability. However, in those studies, blood pressure was increased by approximately 45% in a rapid (30 secs) fashion, and yet only approximately 50% of animals showed increased cerebrovascular permeability (reviewed in ref. 12). In the study of Prado et al[12] the blood pressure increase due to L-NAME was more gradual and of lesser magnitude, and yet 100% of the animals developed vascular abnormalities, suggesting that the pressure increase may not be entirely responsible for the enhanced cerebrovascular permeability.

An L-NAME-induced increase in microvascular permeability or at least transvascular fluid and protein flux has been noted in many but not all organ systems. For example, enhanced protein extravasation (tissue blue dye accumulation) was detected in the stomach, liver, spleen, pancreas, kidney and duodenum.[13] Moreover, some but not all parts of the lung displayed increased leakage to albumin. For example, albumin accumulation was detected in the trachea and bronchus but not in the pulmonary parenchyma. The skin vasculature was also relatively insensitive to L-NAME. Additionally, different organs displayed increased albumin accumulation at different concentrations of L-NAME; evidence of increased albumin accumulation was evident at 0.125 mg/kg in the stomach whereas increased albumin accumulation in the spleen was only apparent at 2 mg/kg in the same study.[13] To test the possibility that increased blood pressure was responsible for the plasma protein accumulation in the extravascular space these authors increased blood pressure with noradrenaline to similar levels as L-NAME and demonstrated that only in the pulmonary circulation was there a notable increase in protein extravasation. These data suggested that increased blood pressure may play a role in the L-NAME-induced rise in protein accumulation in the lung but not other vascular beds.

Another study from the same group demonstrated that L-NAME (2 mg/kg) evoked a 2-fold increase in Evans blue dye leakage within the coronary microvasculature.[14] Maintenance of mean arterial blood pressure at the level observed following L-NAME administration by infusion of noradrenalin did not induce significant protein extravasation in the coronary circulation.[14] These authors concluded that NO may be an important regulator of vascular permeability under physiologic and pathologic conditions. Unlike cardiac muscle vasculature, skeletal muscle microcirculation was unremarkable in its response to L-NAME.[14]

Recently, a novel approach was used to examine microvascular alterations in subepithelial areas of guinea pig airways.[15] These colleagues used 5 nm gold particles that were unable to pass paracellularly under normal conditions. They were able to demonstrate the exudation of gold particles into the lamina propria following topical application of L-NAME. Endothelial retraction is the only way particles of this magnitude could leave the microcirculation without frank hemorrhage which was not observed in that study. This work suggests that inhibition of NO synthase did invoke an increase in microvascular permeability in the airway vasculature. These results were further confirmed by the appearance of ^{125}I-albumin in the lumen of the airways (following lavage) suggesting an increase in microvascular as well as mucosal permeability. In this study, total tissue content of ^{125}I-albumin did not increase after topical L-NAME administration despite the profound increase in microvascular permeability to 5 nm gold particles. A potential explanation proposed by the authors is that L-NAME produces such a marked vasoconstriction that the intravascular pool of ^{125}I-albumin was reduced as much as or more than the extravascular tissue pool of accumulating ^{125}I-albumin. Clearly, simply measuring ^{125}I-albumin accumulation in a tissue may be very crude and misleading with respect to the actual changes at the microvascular barrier.

IS THERE A ROLE FOR cGMP?

It has been well established that nitric oxide synthesis inhibition diminishes cyclic GMP content in endothelial cells as well as other cell types.[16] It is therefore conceivable that the increase in microvascular permeability associated with inhibition of nitric oxide production is a direct reflection of reduced levels of intracellular cyclic GMP. It is well known that increasing intracellular cyclic GMP activates cyclic GMP-dependent protein kinase,[17,18] decreases phosphorylation of myosin light chain[17,18] and induces endothelial cell relaxation, i.e., reduction in the size of interendothelial cell junctions.[3,19-21] It follows that a reduction in cyclic GMP would cause endothelial cell contraction which would produce larger interendothelial junctions, an event that results in a leakier microvascular barrier.[20] The plant cytokine phalloidin which leads to the assembly (stabilization) of the cell cytoskeleton partially reduces L-NAME-induced microvascular alterations,[22] supporting a role for active modification of the endothelial cytoskeleton and endothelial cell retraction with nitric oxide synthesis inhibition. Light and electron microscopic studies may provide important information regarding cytoskeletal rearrangements associated with nitric oxide synthesis inhibition.

To confirm that diminished cGMP levels underlie the L-NAME-induced increase in transvascular fluid and protein flux, tissue cGMP levels were increased directly with 8-bromo-cyclic GMP which bypasses the guanylate cyclase system but replenishes intracellular cyclic GMP content.[19] This manipulation prevented the L-NAME induced rise in transvascular fluid and protein flux (Fig. 4.3). Moreover, an inhibitor of cytosolic guanylate cyclase activation, methylene blue, moderately increased transvascular fluid and protein flux. Similar experiments were subsequently performed in single venules, and 8-bromo-cGMP reversed the L-NAME-induced increase in FITC-albumin out of the postcapillary vessels under study.[22] Collectively, these data strongly support the view that depletion of intracellular cyclic GMP plays an integral role in the increased transvascular fluid and protein flux associated with nitric oxide synthesis inhibition. A note of caution should be added; the effect of 8-bromo-cyclic GMP on L-NAME-induced decrease in the reflection coefficient (σ_d) remains undetermined. However, based on the aforementioned protein flux measurements with 8-bromo-cGMP, and the fact that nitroprusside, which elevates cellular cGMP, reversed the L-NAME-associated decrease in σ_d, these data support the view that the L-NAME-induced rise in microvascular permeability is mediated by cGMP.

To determine whether the L-NAME-induced increase in transvascular fluid and protein flux was inhibitable by any enhancer of cyclic GMP activity, we exposed the intestinal vasculature to

ANF a nitric oxide-independent activator of membrane-associated guanylate cyclase.[19,23,24] Although our initial hypothesis was that any activator of cyclic GMP levels would reverse the effect of L-NAME, there was a significant augmentation of protein and fluid leakage out of the vasculature when L-NAME and ANF were co-infused (Fig. 4.3). These data raised somewhat of a controversy inasmuch as NO donors and 8-bromo-cGMP (both agents increase cGMP) reduced an L-NAME-induced rise in microvascular protein clearance, whereas ANF, which apparently also enhances cellular cGMP, increased microvascular protein clearance. Moreover, the ANF data were consistent with previous reports by Meyer and Huxley[25] wherein ANF caused almost a three-fold increase in baseline transvascular fluid and protein flux in frog mesenteric microvessels. However, when feline intestinal tissue was exposed to ANF, there was no increase in tissue cGMP levels (Fig. 4.4) despite the fact that 1) intestinal cGMP concentrations could be increased with other agents including NO donors and 2) ANF

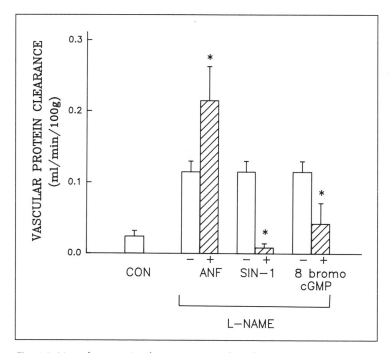

Fig. 4.3. Vascular protein clearance is significantly increased from control (con) values with L-NAME alone. When L-NAME was coinfused with ANF vascular protein clearance increased whereas SIN-1 and 8-bromo-cGMP both decreased the L-NAME-induced increase in vascular protein clearance. *p < 0.05 relative to L-NAME value alone.

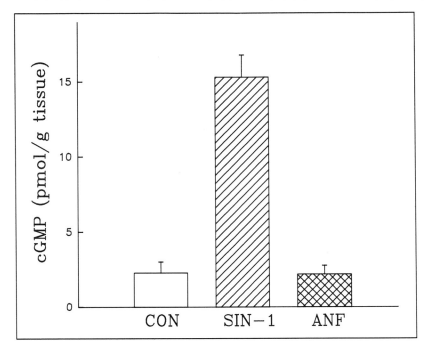

Fig. 4.4. Levels of cGMP in feline intestinal tissue exposed to SIN-1 or ANF. Only SIN-1 increased cGMP levels in the intestine.

increased cGMP in other feline tissues including the carotid artery. These data suggest that the actions of ANF in the intestinal vasculature may be independent of the second messenger cGMP at least in part resolving this controversy.

IS THERE A ROLE FOR LEUKOCYTES?

A technically elaborate study has recently raised some questions about the direct effect of nitric oxide synthesis inhibition on microvascular permeability alterations. Yuan and colleagues[26] used isolated coronary venules (30-70 μm diameter) to determine the effect of nitric oxide synthesis inhibition on changes in microvascular permeability. They observed that as they increased flow of a physiologic salt solution (containing albumin but devoid of blood cells) through postcapillary venules, the vascular permeability to albumin increased. The nitric oxide synthase inhibitor (L-NMMA) decreased baseline and prevented flow-dependent increases in permeability. They concluded that flow continuously modulates permeability of coronary exchange vessels via the production of nitric oxide. It is difficult to reconcile these results with the aforementioned increases in microvascular permeability associated with nitric oxide synthase inhibitors.[8] One possibility may be the lack of leukocytes in the isolated coronary vessel preparations. Leukocytes are known to adhere and emigrate with L-NAME (see chapter 2), and this event is often closely associated with increased microvascular permeability. Therefore, one could envision that in the absence of leukocytes, nitric oxide synthesis inhibition would reduce microvascular permeability, whereas the presence of leukocytes masks this effect and actually enhances microvascular permeability alterations.

Since NO synthesis inhibition causes leukocyte adhesion and emigration,[27] and the latter has previously been shown to cause enhanced microvascular permeability and edema formation,[28,29] then the possibility exists that L-NAME causes a neutrophil-dependent microvascular dysfunction. This hypothesis was been assessed both at the whole organ level by measuring fluid and protein movement into lymphatic vessels as well as in single postcapillary venules by measuring the movement of fluorescently labeled albumin (FITC-albumin) out of these vessels. Local L-NAME administration into the intestinal microvasculature consistently caused an initial rapid rise in vascular protein leakage that persisted for the next 90 mins (Fig. 4.5). In animals that were pretreated with a monoclonal antibody (anti-CD18) that entirely prevented leukocyte adhesion, an unmistakable rise in protein and fluid out of the microvasculature was again observed.[8] In fact the magnitude of the rise in the protein leakage at 15 mins of L-NAME infusion in the presence of the anti-CD18 antibody was not different from that observed with L-NAME alone. In these animals that were pretreated with the anti-CD18 antibody, the vascular protein leakage began to fall so that at 30-60 mins of L-NAME administration, the enhanced vascular protein leakage was entirely prevented by the antiadhesion therapy (Fig. 4.5). These data would suggest that L-NAME causes an early leukocyte-independent and a later leukocyte-dependent rise in microvascular permeability.

Kurose et al[22] confirmed these observations using a model of rat mesentery which allowed for simultaneous measurement of leukocyte adherence as well as albumin leakage in discrete segments of postcapillary venules. Exposure of a venule to L-NAME (superfusion) caused a rapid rise in FITC-albumin leakage (first 10 mins) which reached a plateau for the subsequent 10 mins

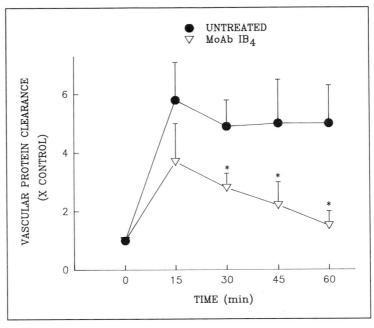

Fig. 4.5. L-NAME infusion into the feline intestinal vasculature increased vascular protein clearance regardless of whether the animals were treated with MoAb IB₄. However, with time in the anti-CD18 antibody-treated group vascular protein leakage began to decline towards control levels. *p < 0.05 relative to respective untreated value.

indicating no further increase in protein movement out of the venule under study. However, a second phase of L-NAME-induced FITC-albumin leakage was observed at approximately 30 mins of L-NAME superfusion. When this profile of FITC-albumin movement out of the vasculature was compared to leukocyte adhesion and emigration within the same venule, two features were observed. The early rise in albumin leakage preceded the L-NAME-induced leukocyte adherence and emigration, whereas the later increase in FITC-albumin correlated closely with the degree of leukocyte adhesion and emigration in the same segment of venule. Kurose et al[22] demonstrated that the later phase of enhanced albumin leakage, but not the earlier phase, was inhibitable with an anti-CD18 antibody. This is consistent with a leukocyte-independent and a leukocyte-dependent increase in albumin leakage.

The importance of leukocytes in the vascular protein leakage out of single postcapillary venules was further characterized by demonstrating that monoclonal antibodies directed against the endothelial ligand for CD18, intracellular adhesion molecule-1

(ICAM-1), was also effective at attenuating the increased vascular albumin leakage induced by L-NAME.[22] Much like the anti-CD18 antibody, the anti-ICAM-1 intervention only attenuated the leukocyte-dependent phase of L-NAME-induced rise in albumin leakage. The early leukocyte-independent phase was not affected by the anti-ICAM-1 antibody. Additionally, a monoclonal antibody directed against P-selectin, an adhesion molecule responsible for leukocyte rolling (prerequisite for leukocyte adhesion and emigration), also prevented the second phase of L-NAME-induced rise in albumin leakage. This observation further supports the importance of leukocytes in the later rise in albumin leakage out of postcapillary venules. The P-selectin data must be interpreted with caution inasmuch as P-selectin is also found on platelets, and L-NAME caused platelet-leukocyte aggregates that were inhibited by the P-selectin antibody.[22] It is therefore conceivable that some of the protection in reducing albumin leakage by the anti-P-selectin antibody may be related to the possibility that platelet-leukocyte aggregates release factors that either directly or indirectly affect the vascular alterations.

These single venule studies are consistent with the whole organ data that inhibition of nitric oxide synthesis leads to a rapid increase in vascular protein leakage that is initially leukocyte-independent and subsequently leukocyte dependent.[30] These data however do not entirely explain observations by Yuan et al[26] who observed that L-NAME causes a decrease in vascular permeability in the absence of blood cells in single venules. Whether different tissues (coronary vs mesentery) and different species (pig vs cat or rat) explain this discrepancy awaits future experiments.

DOES EXOGENOUS NITRIC OXIDE REDUCE BASELINE PERMEABILITY?

Until now we have discussed the effects of removing endogenous nitric oxide from the vasculature, and the data with some exceptions would support the view that endogenous nitric oxide production is a homeostatic regulator of vascular permeability, and reductions in synthesis of this autacoid can increase the movement of albumin across the endothelial barrier at least in part due to adhering leukocytes. The possibility that enhancing nitric oxide levels could reduce baseline permeability characteristics will be discussed in this section.

Oliver[31] recently tested the possibility that exogenous NO could modulate endothelial permeability. The flux of ^{14}C-sucrose (a marker of the paracellular pathway) was examined across bovine aortic endothelial cells grown on collagen-coated filters. This investigator demonstrated that the nitric oxide donor, glyceryl trinitrate as well as the cGMP analog 8-bromo-cGMP decreased baseline permeability across arterial endothelial monolayers in vitro. Depletion of L-arginine, the precursor for NO, from monolayers of arterial endothelium, increased baseline permeability to ^{14}C-sucrose by 20% in endothelial monolayers. Moreover, replenishing L-arginine depleted monolayers with L-arginine but not D-arginine significantly decreased baseline permeability. Similar results have been obtained using human umbilical vein endothelial monolayers. Yamada et al[32] demonstrated that sodium nitroprusside, 8-bromo-cGMP or dibutyrl cGMP which increase intracellular cGMP concentrations lowered albumin transfer and increased electrical resistance. Taken together, these data implicate a role for the L-arginine-NO-cGMP pathway in the maintenance of endothelial permeability. Moreover, the study by Yamada et al suggests that in vitro, increasing NO and/or cGMP can enhance the baseline restrictive properties of endothelial monolayers. Whether these studies can be extended to the in vivo condition is unclear. For example, the endothelial monolayers were used in experiments 1 day after plating, raising the possibility that baseline permeability was artificially high, and perhaps at 2 or 3 days following plating the baseline may have been lower and unresponsive to NO donors.

Local infusion of NO donors into the intestinal circulation with CAS 754 or SIN-1, dramatically reduced baseline vascular protein flux out of the intestinal microvasculature.[9] Similar changes in baseline vascular protein flux were observed when intracellular levels of cyclic GMP were increased with 8-bromo-cGMP. The changes in vascular protein clearance observed in that study were by no means trivial inasmuch as increasing intracellular cyclic GMP levels reduced baseline vascular protein clearance by 75-90%. Although it is tempting to translate these changes in vascular protein clearance into a reduction in baseline microvascular permeability this assumption has proved to be incorrect. Recently, we infused an NO donor (CAS 754) into the intestinal microcirculation and simultaneously modulated capillary hydrostatic pressure by locally infusing epinephrine into the intestinal circulation. We observed (Fig. 4.6) that if capillary hydrostatic pressure was increased in this preparation, the transvascular fluid and protein flux rapidly rose in the presence of CAS 754. Clearly, depending on the driving force (P_{CAP}), a conclusion could be reached that nitric oxide donors either decrease, increase or do not change transvascular protein flux.

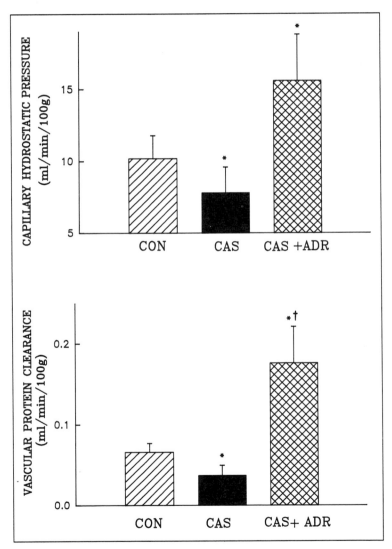

Fig. 4.6. *Capillary hydrostatic pressure and vascular protein clearance were measured in the same experiments, and the data reveal that CAS 754, an NO donor, decreased both capillary hydrostatic pressure as well as vascular protein clearance. When the capillary hydrostatic pressure was increased with adrenaline the vascular protein clearance rose in parallel. *p < 0.05 relative to control. † p < 0.05 relative to CAS value.*

Since it is difficult to achieve steady-state conditions with adrenalin infusion, in another series of experiments, we measured σ_d at increasing concentrations of CAS 754 infused locally into the small intestinal circulation (Fig. 4.7). At filtration rate-independent conditions (eliminates P_{CAP} as a confounding variable) where the true sieving properties of the vasculature (microvascular permeability) are evident, we observed that the nitric oxide donor CAS 754 did not affect σ_d. These data suggest that sufficient amounts of endogenous NO are produced to maintain a tight endothelial barrier and that supplementing the system with excess exogenous NO does not further reduce microvascular permeability. Although these studies are at odds with the aforementioned in vitro data wherein NO donors did reduce the permeability to albumin, extending in vitro data to the in vivo conditions can be problematic because of the potential complications of removing the endothelium from the vasculature and then growing the cells in culture. Moreover, the importance of continuous shear and hydrostatic pressure on the porosity characteristics of endothelium in

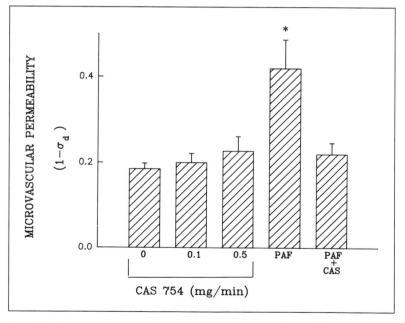

Fig. 4.7. Microvascular permeability was measured as $1\text{-}\sigma_d$. CAS 754 infused continuously at very high concentrations did not affect baseline microvascular permeability. PAF however increases $1\text{-}\sigma_d$ suggesting a rise in microvascular permeability. Coinfusion of PAF + CAS 754 prevented the rise in PAF-induced increase in $1\text{-}\sigma_d$. *$p < 0.05$ relative to control value (0 mg/min).

vivo versus the lack thereof in vitro, makes it difficult to compare the two systems.

Interestingly, Meyer and Huxley recently reported that the nitrovasodilator, nitroprusside, increased hydraulic conductivity, a measure of capillary exchange capacity, in frog mesenteric capillaries.[25] A capillary was cannulated and perfused with frog-Ringer's perfusate containing 10 mg/ml dialyzed BSA. These authors observed a reversible 2.5-fold increase in hydraulic conductivity with sodium nitroprusside, suggesting that exogenous NO may enhance transendothelial albumin movement. These authors point out however that the increase observed with nitroprusside is distinct in magnitude from the gross morphological changes including formation of gap junctions that are observed with histamine and other modulators of permeability. Moreover most gap junctions occur in venules and rarely in capillaries. Since the increase in hydraulic conductivity in capillaries occurs independent of gap formation, it likely involves a different set of mechanisms than what might be observed in intact postcapillary venules. Nevertheless, it raises the possibility that NO may elicit different responses in different segments of the microvasculature.

INFLAMMATION-INDUCED MICROVASCULAR PERMEABILITY AND THE ROLE OF NITRIC OXIDE

One of the most controversial areas of research regarding NO is whether this autacoid is responsible for or inhibits the permeability changes invoked by various proinflammatory mediators. One complication in interpreting many of the studies that have attempted to examine this hypothesis is the mode of measuring the leakage of albumin across the vasculature. Often potentially large changes in microvascular pressures and substantial changes in blood flow and fluid delivery to a tissue are ignored. Measures of fluid accumulation (edema formation) as an index of increased microvascular permeability may or may not be representative of permeability changes. Finally, the role of the lymphatic system which continuously removes fluid and protein from tissue is rarely considered when assessing edema. Alterations in the efficiency of the lymphatic pump as a result of NO administration or removal could influence edema formation. Therefore, the role of NO in stimulated states of microvascular permeability needs to be interpreted with great caution.

Table 4.1 summarizes some of the studies that have examined the effect of NO donors or NO synthase inhibitors on edema formation, protein accumulation or various indices of microvascular permeability. Yuan et al[33] have recently demonstrated that a histamine-induced rise in venular permeability can be inhibited by both a phospholipase C inhibitor as well as an inhibitor of NO synthase or cyclic GMP. The NO donor, nitroprusside, also increased venular permeability, an effect that was blocked by a guanylate cyclase inhibitor but not by the phospholipase C inhibitor. The authors proposed that the NO stimulated guanylate cyclase pathway is the final signal in the histamine-induced phospholipase C activation process leading to increased venular permeability. This hypothesis is supported by the recent observation that histamine binds to the H_1-receptor and increases cGMP levels through a phospholipase C-dependent enzymatic conversion of L-arginine to NO. These observations are also in part consistent with a more recent study which demonstrated that histamine-induced albumin accumulation in skin was reduced by L-NAME.[34] However, similar reductions were observed with L-arginine, and L-arginine did not reverse the effects of L-NAME. These results are difficult to interpret and raise some questions regarding the possibility that the effects of L-NAME and L-arginine in the presence of histamine were unrelated to NO production.

In addition to histamine, NO synthesis inhibition also prevents substance P, bradykinin, ADP and carrageenin-induced rise in permeability (Table 4.1). Carrageenin increased FITC-albumin accumulation at skin sites, an event inhibitable by 35% with L-NAME.[35] Moreover, rat paw edema induced by dextran was inhibited by L-NAME or L-NMMA. Paul et al[34] reported a reduction in bradykinin-induced albumin accumulation in guinea pig skin with L-NAME. Although neither study measured the degree of vasoconstriction to the L-NAME treated skin sites, the latter study demonstrated that a vasoconstrictor phenylephrine (α-adrenoceptor agonist) was as effective as L-NAME in preventing bradykinin associated rise in albumin accumulation. Therefore, profound reductions in blood flow and albumin delivery with L-NAME could explain the inhibitory effect of the NO synthase inhibitor on bradykinin-induced albumin accumulation. Hughes et al[36] demonstrated that L-NAME could reduce substance P

Table 4.1. Mediator-induced vascular protein leakage and the role of NO donors or inhibitors

Model	L-Arginine/ NO donor	NO inhibitor	Reference
Histamine-stimulated albumin leakage in isolated porcine coronary venules	Augmented	Decreased	Yuan et al[33]
Histamine-stimulated albumin accumulation in guinea pig skin	Decreased	Decreased	Paul et al[34]
Bradykinin-stimulated albumin accumulation in guinea pig skin	Augmented	Decreased	Paul et al[34]
Carrageenin-stimulated albumin accumulation in rat skin	Augmented	Decreased	Ialenti et al[35]
Dextran-stimulated rat paw edema	Augmented	Decreased	Ialenti et al[35]
Substance P-stimulated albumin accumulation in rat skin	ND	Decreased	Hughes et al[36]
Polycation-stimulated rat paw edema	ND	Decreased	Antunes et al[38]
Mustard oil-stimulated albumin accumulation in rat skin	ND	No effect	Lippe et al[39]
PAF-stimulated albumin accumulation in numerous rat organs	ND	Augmented	Filep and Filep[14,40]
ET-1-stimulated albumin accumulation in numerous rat organs	ND	Augmented	Filep et al[40]
Vagal-stimulated albumin accumulation in rat pulmonary vasculature	ND	Augmented	Liu et al[41]
PAF-stimulated reflection coefficient in cat intestinal vasculature	Decreased	ND	Kubes et al[42]
Oxidant-stimulated filtration coefficient in rabbit lung	Decreased	No effect	Kavanagh et al[43]
Histamine-stimulated paw edema in mice	Augmented	Decreased	Oyanagui and Sato[44]
Serotonin-stimulated paw edema in mice	Decreased	Augmented	Oyanagui and Sato[44]
Histamine, LTC4, ADP, bradykinin-stimulated albumin leakage in hamster cheek pouch venules	Augmented	Decreased	Mayhan[45]
Reperfusion-stimulated albumin leakage in rat mesentery venules	Decreased	No effect	Kurose et al[47]

ND Not determined

associated ^{125}I-albumin accumulation in skin by 32%. These authors also measured blood flow in this vascular bed and observed a 50% reduction in blood flow to this region when L-NAME was introduced. It was difficult to dissociate the vasoconstrictive effects of L-NAME from the reduction in ^{125}I-albumin accumulation. In fact previous results from this group demonstrated that the substance P-induced rise in protein accumulation was potentiated by vasodilators and inhibited by vasoconstrictors.[37] These observations raise questions about the importance of L-NAME as anything more than simply a vasoconstrictor in these studies.

This is further supported by Antunes et al[38] who reported that polycation-induced edema formation in rat hind-paw was reduced by 50% with the NO synthase inhibitor, L-NMMA. The immediate conclusion might be that endogenous NO is in part responsible for the increase in polycation-induced edema formation and the associated increase in microvascular permeability. However, if a vasodilator without effects on microvascular permeability was injected simultaneously with L-NMMA, the inhibitory effect of L-NMMA was no longer apparent. These authors therefore concluded that the vasoconstrictive effects of L-NMMA were responsible for the reduced edema formation rather than a direct effect on microvascular permeability. Although not tested in this study, local addition of a vasodilator such as nitroprusside could conceivably enhance the delivery of fluid to the inflamed paw and further increase edema formation. Clearly, addition of vasoconstrictive or vasodilatory agents can affect fluid and protein flux across the microvasculature independent of any changes in microvascular permeability.

Lippe et al[39] demonstrated that mustard oil-induced neurogenic inflammation of the rat paw skin was composed of increased albumin accumulation as well as vasodilation. In this study, L-NAME (43 μmol/kg) reduced the cutaneous hyperemia but had no effect on protein or fluid accumulation in the tissue. These data dissociate the effect of NO on increased blood flow and increased albumin accumulation. The rationale for the use of 43 μmol/kg of L-NAME was not provided, however it would have been interesting to determine the biologic effects of higher concentrations of L-NAME to see whether as blood flow was further decreased, albumin accumulation could also be reduced. In addition to this study, there have been a number of other reports that have not been able to invoke a role for NO in increased albumin accumulation. These studies are also summarized in Table 4.1.

Finally, Table 4.1 summarizes data wherein an antipermeability effect for NO has been invoked. For example, L-NAME enhanced platelet-activating factor-induced albumin extravasation in the liver, spleen, kidney, stomach and duodenum.[13] Importantly, noradrenalin did not have similar enhancing properties as L-NAME in these tissues. Similar findings were reported for L-NAME and endothelin-1 in pulmonary parenchyma, stomach, duodenum, pancreas, liver and kidney.[40] Although blood flow was not measured in these experiments, it is difficult to invoke a role for this hemodynamic parameter inasmuch as albumin accumulation increased despite the addition of a vasoconstrictor. Moreover, increased blood pressure with noradrenalin (to the L-NAME level) without affecting albumin leakage was used to counter the argument that the increased albumin accumulation was a result of increased arterial pressure invoked by L-NAME. A counter-argument against this explanation is the possibility that norepinephrine and L-NAME exerted their actions at different sites of the vasculature. For example, if L-NAME caused in addition to vasoconstriction, venoconstriction whereas norepinephrine only modulated the former, then increased capillary hydrostatic pressure might cause protein efflux out of the vasculature with L-NAME. However, there is no evidence to support L-NAME as a potent venoconstrictor, and in fact we have measured a decrease rather than an increase in intestinal capillary hydrostatic pressure with L-NAME.[8] Finally, Liu et al[41] demonstrated that pulmonary plasma leakage was increased following vagal stimulation and L-NAME significantly enhanced plasma leakage to parenchyma in a dose-related and L-arginine-reversible manner. Again, increasing blood pressure with phenylephrine to a similar extent as that with L-NAME did not enhance vagal stimulation-induced protein accumulation in lung tissue.

Addition of exogenous nitric oxide has proven to reduce increases in microvascular permeability associated with platelet-activating factor[42] or biochemical production of oxidants.[43] PAF caused a very profound decrease in σ_d (reflection coefficient) to total plasma proteins suggesting a dramatic increase in microvascular permeability in the feline small intestine (illustrated as $1-\sigma_d$ in Fig. 4.7). However when CAS 754 (NO donor) was coadministered with PAF, the nitric oxide donor significantly reduced the PAF-induced rise in microvascular permeability (Fig. 4.7).[42] The reflection coefficient is measured at filtration-rate independent conditions so that hemodynamic alterations including blood flow, surface area and hydrostatic pressure changes do not affect this index of microvascular permeability. Kavanagh et al[43] measured the lung capillary filtration coefficient (K_{fc}) and observed that inhaled NO reduced the K_{fc} rise caused by infusion of an oxidant-generating system. L-NAME did not exacerbate the oxidant-induced K_{fc}, and therefore the authors concluded that exogenous but not endogenous NO can reduce capillary permeability in a model of oxidant-induced acute lung injury.

Clearly, the role of NO on microvascular permeability in situations where a proinflammatory manipulation is used is presently not understood, and a resolution does not appear readily available. One might argue that many of the apparent permeability measurements are not a true measure of permeability and simply reflect changes in other hemodynamic parameters such as blood flow. However, this does not explain the contradictory results of Yuan et al[33] and others[8,42] wherein permeability measurements were made independent of blood flow or blood pressure changes. Additionally, one might argue that different tissues might respond differently to similar stimuli. In this regard, it is intriguing to see that the skin and joint appear to consistently be sites wherein L-NAME prevents rises in albumin accumulation, whereas the intestine and lung appear to demonstrate opposite effects. Nevertheless, there are reports where different stimulators (histamine vs serotonin) in the same tissue from the same laboratory responded differently to NO donors.[44]

One shortcoming of many of these studies is the different models, different measurements and most importantly the different concentrations of NO synthase inhibitors and NO donors used. For example, L-NMMA at 1 µM has been shown to completely prevent histamine-induced protein leakage in the hamster cheek pouch,[45] however at least 100 µM L-NMMA are required to significantly inhibit (70%) of the enzymatic function of the constitutive NO synthase in the rat mesentery.[46] In fact, 1 µM L-NMMA did not affect NO synthase activity in the rat mesentery. Whether these data can be extrapolated to the hamster cheek pouch remains to be seen, however confirmation that NO synthase activity is indeed blocked with the NO synthase inhibitors is clearly warranted in future studies.

Finally, agonists can increase microvascular permeability alterations by directly activating the endothelium; these include histamine, bradykinin and serotonin. Other mediators likely increase microvascular permeability indirectly via activation of inflammatory cells such as neutrophils which adhere and emigrate out of the vasculature simultaneously increasing microvascular permeability. LTB_4 and perhaps C5a function in this manner. Finally, there is a class mediators that may cause a dual effect (neutrophil-dependent and neutrophil-independent) including PAF, activators of mast cells and various agents that are used to invoke an inflammatory response such as carrageenin. A systematic study to determine whether NO has differing effects on a neutrophil-dependent increase in microvascular permeability versus a neutrophil-independent increase in microvascular permeability remains to be performed. Kurose et al[47] have examined the effect of NO donors on vascular protein leakage in single venules exposed to ischemia/reperfusion. They observed a profound increase in microvascular permeability that was associated with increased leukocyte adhesion. When NO donors were superfused onto these vessels, the adhesion, emigration and microvascular permeability were attenuated. Therefore, in this particular neutrophil-dependent model of inflammation, the reduction in enhanced leukocyte adhesion and microvascular permeability coincided. Whether the same NO donor regimen would affect histamine-induced (neutrophil-independent) rise in permeability remains to be seen.

SUMMARY

In conclusion, this chapter summarizes the data to implicate NO as an endogenous modulator of microvascular permeability. It does not resolve the controversy that seems to be prevalent in terms of whether NO increases or decreases microvascular permeability. It does however highlight some of the shortcomings and areas that require more attention to try and determine the role of NO in the microvasculature. Finally, a more concerted effort is needed to link the role of leukocyte adhesion, microvascular permeability alterations and the effects of NO on these related events in inflammation.

REFERENCES

1. Grega GJ. Role of the endothelial cell in the regulation of microvascular permeability to molecules. Fed Proc 1986; 40:75-76.
2. Grega GJ, Adamski SW, Dobbins DE. Physiological and pharmacological evidence for the regulation of permeability. Fed Proc 1986; 45:96-100.
3. Svensjo E, Grega GJ. Evidence for endothelial cell-mediated regulation of macromolecular permeability by postcapillary venules. Fed Proc 1986; 45:89-95.
4. Majno G, Shea SM, Leventhal M. Endothelial contraction produced by histamine-type mediators. J Cell Biol 1969; 42:647-672.
5. Amelang E, Prasad CM, Raymond RM et al. Interactions among inflammatory mediators on edema formation in the canine forelimb. Circ Res 1981; 49:298-306.
6. Anderson GL, Miller FN, Xiu R-J. Inhibition of histamine-induced protein leakage in rat skeletal muscle by blockade of prostaglandin synthesis. Microvasc Res 1984; 28:51-61.
7. Grega GJ, Maciejko JJ, Raymond RM et al. The interrelationship among histamine, various vasoactive substances, and macromolecular permeability in the canine forelimb. Circ Res 1980; 46:264-275.
8. Kubes P, Granger DN. Nitric oxide modulates microvascular permeability. Am J Physiol 1992; 262:H611-H615.
9. Kubes P. Nitric oxide-induced microvascular permeability alterations: a regulatory role for cGMP. Am J Physiol 1993; 265:H1909-H1915.
10. Taylor AE, Granger DN. Exchange of macromolecules across the microcirculation. In: Handbook of Physiology. 1983:467-520.
11. Granger DN, Benoit JN, Suzuki M et al. Leukocyte adherence to venular endothelium during ischemia-reperfusion. Am J Physiol 1989; 257:G683-G688.
12. Prado R, Watson BD, Kuluz J et al. Endothelium-derived nitric oxide synthase inhibition. Effects on cerebral blood flow, pial artery diameter, and vascular morphology in rats. Stroke 1992; 23:1118-1124.
13. Filep JG, Foldes-Filep E. Modulation by nitric oxide of platelet-activating factor-induced albumin extravasation in the conscious rat. Br J Pharmacol 1993; 110:1347-1352.
14. Filep JG, Foldes-Filep E, Sirois P. Nitric oxide modulates vascular permeability in the rat coronary circulation. Br J Pharmacol 1993; 108:323-326.
15. Erjefalt JS, Erjefalt I, Sundler F et al. Mucosal nitric oxide may tonically suppress airways plasma exudation. Am J Respir Crit Care Med 1994; 150:227-232.
16. Moncada S. The L-arginine: Nitric oxide pathway. Acta Physiol Scand 1992; 145:201-227.
17. Rapoport RM, Draznin MB, Murad F. Endothelium-dependent relaxation in rat aorta may be mediated through cyclic GMP-dependent

protein phosphorylation. Nature 1983; 306:174-176.

18. Rapoport RM, Draznin MB, Murad F. Sodium nitroprusside-induced protein phosphorylation in intract rat aorta is mimicked by 8-bromo cyclic GMP. Proc Natl Acad Sci USA 1982; 79:6470-6474.

19. Goy MF. cGMP: the wayward child of the cyclic nucleotide family. TINS 1991; 14:293-299.

20. Miller FN, Sims DE. Contractile elements in the regulation of macromolecular permeability. Fed Proc 1986; 45:84-88.

21. Wysolmerski RB, Lagunoff D. Involvement of light-chain kinase in endothelial cell retraction. Proc Natl Acad Sci USA 1990; 87:16-20.

22. Kurose I, Kubes P, Wolf R et al. Inhibition of nitric oxide production: Mechanisms of vascular albumin leakage. Circ Res 1993; 73:164-171.

23. Koesling D, Bhome E, Schultz G. Guanylyl cyclases, a growing family of signal-transducing enzymes. FASEB J 1991; 5:2785-2791.

24. Waldman SA, Murad F. Cyclic GMP synthesis and function. Pharmacol Rev 1988; 39:163-196.

25. Meyer DJ, Huxley VH. Capillary hydraulic conductivity is elevated by cGMP-dependent vasodilators. Circ Res 1992; 70:382-391.

26. Yuan Y, Granger HJ, Zawieja DC et al. Flow modulates coronary venular permeability by a nitric oxide-related mechanism. Am J Physiol 1992; 263:H641-H646.

27. Kubes P, Suzuki M, Granger DN. Nitric oxide: An endogenous modulator of leukocyte adhesion. Proc Natl Acad Sci USA 1991; 88:4651-4655.

28. Wedmore CV, Williams TJ. Control of vascular permeability by polymorphonuclear leukocytes in inflammation. Nature 1981; 289:646-650.

29. Kubes P, Grisham MB, Barrowman JA et al. Leukocyte-induced vascular protein leakage in cat mesentery. Am J Physiol 1991; 261:H1872-H1879.

30. Niu X-F, Smith CW, Kubes P. Intracellular oxidative stress induced by nitric oxide synthesis inhibition increases endothelial cell adhesion to neutrophils. Circ Res 1994; 74:1133-1140.

31. Oliver JA. Endothelium-derived relaxing factor contributes to the regulation of endothelial permeability. J Cell Physiol 1992; 151:506-511.

32. Yamada Y, Furumichi T, Furui H et al. Roles of calcium, cyclic nucleotides, and protein kinase C in regulation of endothelial permeability. Arteriosclerosis 1990; 10:410-420.

33. Yuan Y, Granger HJ, Zawieja DC et al. Histamine increases venular permeability via a phospholipase C-NO synthase-guanylate cyclase cascade. Am J Physiol 1993; 264:H1734-H1739.

34. Paul W, Douglas GJ, Lawrence L et al. Cutaneous permeability responses to bradykinin and histamine in the guinea-pig: possible differences in their mechanism of action. Br J Pharmacol 1994; 111:159-164.

35. Ialenti A, Ianaro A, Moncada S et al. Modulation of acute inflammation by endogenous nitric oxide. Eur J Pharmacol 1992; 211:177-182.
36. Hughes SR, Williams TJ, Brain SD. Evidence that endogenous nitric oxide modulates oedema formation induced by substance P. Eur J Pharmacol 1990; 191:481-484.
37. Brain SD, Williams TJ. Interactions between the tachykinins and calcitonin gene-related peptide lead to modulation of oedema formation and blood flow in rat skin. Br J Pharmacol 1989; 97:77-82.
38. Antunes E, Mariano M, Cirino G et al. Pharmacological characterization of polycation-induced rat hind-paw oedema. Br J Pharmacol 1990; 101:986-990.
39. Lippe IT, Stabentheiner A, Holzer P. Participation of nitric oxide in the mustard oil-induced neurogenic inflammation of the rat paw skin. Eur J Pharmacol 1993; 232:113-120.
40. Filep JG, Földes-Filep E, Rousseau A et al. Vascular responses to endothelin-1 following inhibition of nitric oxide synthesis in the conscious rat. Br J Pharmacol 1993; 110:1213-1221.
41. Liu S, Kuo H-P, Sheppard MN et al. Vagal stimulation induces increased pulmonary vascular permeability in guinea pig. Am J Respir Crit Care Med 1994; 149:744-750.
42. Kubes P, Reinhardt PH, Payne D et al. Excess nitric oxide does not cause cellular, vascular or mucosal dysfunction in the cat small intestine. Am J Physiol 1995; (in press)
43. Kavanagh BP, Mouchawar A, Goldsmith J et al. Effects of inhaled NO and inhibition of endogenous NO synthesis in oxidant-induced acute lung injury. J Appl Physiol 1994; 76:1324-1329.
44. Oyanagui Y, Sato S. Histamine paw edema of mice was increased and became H_2-antagonist sensitive by co-injection of nitric oxide forming agents, but serotonin paw edema was decreased. Life Sciences 1993; 52:159-164.
45. Mayhan WG. Nitric oxide accounts for histamine-induced increases in macromolecular extravasation. Am J Physiol 1994; 266:H2369-H2373.
46. Kurose I, Wolf R, Grisham MB et al. Microvascular responses to inhibition of nitric oxide production: role of active oxidants. Circ Res 1995; 76:30-39.
47. Kurose I, Wolf R, Grisham MB et al. Modulation of ischemia/reperfusion-induced microvascular dysfunction by nitric oxide. Circ Res 1994; 74:376-382.

===== CHAPTER 5 =====

Nitric Oxide Modulates Leukocyte Function in Ischemia/Reperfusion

Paul Kubes

INTRODUCTION

There is much evidence to implicate infiltrating leukocytes as primary mediators of ischemia/reperfusion-induced organ dysfunction. Intravital microscopy experiments have revealed that immediately upon the onset of reperfusion, leukocytes infiltrate afflicted tissue (see Figs. 2.2A and B, chapter 2). Moreover, depletion of circulating leukocytes with antineutrophil serum significantly attenuated reperfusion-induced tissue injury. Finally, preventing leukocyte infiltration into postischemic tissues with monoclonal antibodies that immunoneutralize the CD18 (β_2-integrin) adhesive glycoprotein complex dramatically attenuate reperfusion injury supporting the view that the adhesive mechanism is critical to leukocyte infiltration and subsequent impairment of tissue. Table 5.1 summarizes studies that have described the different mechanisms underlying leukocyte recruitment into postischemic tissue. The mechanisms include increased P-selectin-dependent leukocyte rolling and PAF and LTB_4-dependent, CD18-associated leukocyte adhesion. Additionally, there is much evidence to suggest a role for various oxidants including superoxide and H_2O_2 as well as mast cell degranulation in the process leading to leukocyte recruitment

Nitric Oxide: A Modulator of Cell-Cell Interactions in the Microcirculation, edited by Paul Kubes. © 1995 R.G. Landes Company.

Table 5.1. Ischemia/reperfusion and nitric oxide synthesis inhibition: similar mechanisms of leukocyte recruitment

Mechanism	Ischemia/Reperfusion	L-Name
P-Selectin	References	References
1) Increased expression	45,71	72
2) Increased rolling	7,18	72
Adhesion		
1) CD18-dependent	18,73	11,52
2) PAF-dependent	43	74
3) LTB$_4$-dependent	42,75	74
Oxidants		
1) Increased production	76,77	78,79
2) Cause adhesion	33,73	38
Mast cells		
1) Degranulation	13,80	38,39
2) Cause adhesion	7	38

into postischemic tissue (see references for review). Also shown in Table 5.1 are the remarkably similar series of events leading to leukocyte recruitment observed following the inhibition of nitric oxide synthesis (see chapter 2 for more detail). From these studies it is tempting to conclude that inhibition of nitric oxide synthesis may be an important component in ischemia/reperfusion-induced leukocyte recruitment. This chapter will summarize data regarding the role of nitric oxide as a potential mediator of ischemia/reperfusion-induced leukocyte recruitment.

NITRIC OXIDE LEVELS ARE REDUCED IN ISCHEMIA/REPERFUSION

The concept that nitric oxide-dependent biologic responses are reduced during reperfusion of ischemic tissue has been well documented in various tissues[1-3] as well as isolated cells.[4] The vasodilator response to acetylcholine (a nitric oxide dependent event) has been reported to be depressed by 50% at 2.5 mins of reperfusion.[3] Longer periods of reperfusion (5 mins, 20 mins, 180 mins, 270 mins) translated into more pronounced decrements in response to acetylcholine. Acetylcholine added to vascular rings exposed only to ischemia without reperfusion gave normal relaxation responses suggesting that the impairment occurs during the reperfusion phase of ischemia/reperfusion (I/R). The vasodilator response to a non-receptor nitric oxide-dependent vasodilator (A23187) was also impaired but not before 20 mins of reperfusion. This 2.5 vs 20 mins discrepancy between acetylcholine and A23187 may suggest some impairment in 1) the functionality of the acetylcholine receptor or 2) the transduction of a signal for the endothelium to produce nitric oxide. Relaxation to an endothelium-independent dilator $(NaNO_2)$ was preserved during reperfusion suggesting that the impairment was associated with the endothelium rather than at the level of the smooth muscle. These observations have now been confirmed in other postischemic vascular beds including skeletal, cerebral and mesenteric circulations.[1,5,6]

Superoxide dismutase abolished the reperfusion-induced impairment in vasorelaxation to acetylcholine or A23187, suggesting a role for superoxide or a superoxide-derived oxidant as the inhibitor of coronary vessel relaxation during reperfusion.[3] Although the source of superoxide was not determined, neutrophils, monocytes and endothelium can all produce this reactive oxygen intermediate, a molecule known to potently inhibit nitric oxide. However, the fact that myeloperoxidase levels (an index of neutrophil accumulation) did not increase until 180 and 270 mins of reperfusion[3] suggested that neutrophils were unlikely to be involved in the 2.5 mins reperfusion-induced endothelium dysfunction. Although the authors postulated that perhaps non-neutrophil sources such as endothelium was a potential candidate, this hypothesis remains untested in the feline myocardium.

A potential problem with measuring MPO levels to determine leukocyte recruitment is the low level of sensitivity. This assay therefore requires a large degree of leukocyte recruitment before significant accumulation is realized. Moreover, significant adhesion of neutrophils within the vasculature without emigration will also likely remain unnoticed using MPO measurements. Of course it is essentially impossible to examine the microcirculation of the myocardium on line, however visualization of the postischemic feline vasculature in the mesentery can be accomplished. This technique revealed profound leukocyte influx (rolling, adhesion and emigration) as early as 2 mins following reperfusion,[7] and this event persisted for at least the next 2 hrs. It is conceivable that this time course also occurs within the myocardial vasculature, however as already mentioned, the MPO assay may not be sensitive enough to detect the early leukocyte influx. These data suggest that the endothelium could be exposed to high levels of leukocytes very early into the reperfusion phase. These data however do not establish a role for these cells in the early oxidant-dependent endothelial dysfunction.

Convincing data to suggest that neutrophils were not contributing to the oxidative stress that inhibits nitric oxide production in the early part of reperfusion were provided by Lefer and Aoki.[8] These investigators demonstrated that saline-perfused, isolated rat hearts with undetectable levels of MPO activity still manifested an impairment in vasodilation in response to acetylcholine, and this dysfunction was inhibited with SOD. These data strongly support the view that the initial superoxide-mediated impairment in relaxation is unrelated to leukocytes. However these data seemed to be species specific inasmuch as these authors subsequently demonstrated that isolated cat hearts exposed to 90 mins of low-flow perfusion and 20 mins of reperfusion with a blood cell-free solution exhibited only a 20-25% reduction in endothelial function.[9] Exposing these vascular rings to activated neutrophils further enhanced endothelial dysfunction in the cat coronary artery rings. It was also determined that much of the impairment of vasorelaxation was mediated by the release of superoxide from the adherent neutrophils. Both antiadhesion and antioxidant therapy reduced the vasoconstriction. Addition of an inhibitor of nitric oxide synthesis completely prevented the vasoconstrictive effect of the neutrophils leading the authors to conclude that the phagocytes were impairing vasorelaxation by inactivating nitric oxide released by the endothelium.

A qualification should be made; the leukocyte-endothelial cell interaction seen within the postischemic vasculature has to date been observed on the venular side of the microcirculation, and therefore, the resistance vessels regulating dilation would not be exposed to rolling or adhering leukocytes at any time during reperfusion. One possible explanation has been that the shear forces pushing leukocytes along the length of vessels is too high to allow leukocytes to adhere on the arterial side. However, the coronary circulation has cyclical changes in shear (from low to very high) with each beat and may be more likely to support leukocyte-endothelial cell interactions on the arterial side of the vasculature. Indeed, Albertine et al[10] recently reported a doubling of leukocytes in precapillary arterioles during I/R without a change in circulating leukocyte numbers suggesting an accumulation of leukocytes in these vessels.

Based on these data, and the previous observation that inhibition of nitric oxide synthesis under normal conditions leads to increased leukocyte adhesion,[11] it is conceivable that diminished nitric oxide levels during the first few minutes of reperfusion could increase leukocyte adhesion causing further endothelial-derived dysfunction. A direct test of this hypothesis is difficult, however there is some indirect evidence to support this view. Ma et al[12] demonstrated that the diminished nitric oxide production observed in postischemic vessels correlated closely with increased leukocyte adhesion. When postischemic coronary arteries were treated with L-arginine, the precursor of nitric oxide, subsequent vasorelaxation to acetylcholine was restored suggesting that nitric oxide levels were returned towards control. Under these conditions, the increased leukocyte adhesion during ischemia/reperfusion was greatly diminished. These authors concluded that the reduced nitric oxide release during reperfusion was a determining factor in the increased neutrophil adhesion.

It is most interesting that there is a notable reduction of nitric oxide levels in the postischemic feline vasculature at approximately 2-3 mins of reperfusion and that in our hands, there is a tremendous hyperemic response in single postcapillary venules for the first 1-2 minutes followed by a profound influx in leukocytes within the next few minutes. Although purely speculative, one could envision a scenario wherein nitric oxide is increased above baseline within the first couple of minutes of reperfusion (Fig. 5.1A) followed by a significant decrease in nitric oxide production (and blood flow), an event closely associated with progressively increased leukocyte rolling and adhesion (Figs. 5.1B and C). The sequence of events is important inasmuch as it may answer the question of whether the reduction in nitric oxide is responsible for subsequent leukocyte adhesion or whether the adhering leukocytes reduce nitric oxide production. Until such time as an online in vivo measure of nitric oxide is available this issue will be difficult to resolve.

In another study, Kurose et al[13] measured nitrate and nitrite concentrations (end products of nitric oxide metabolism) of arterial and venular (draining the rat intestine) blood prior to and following reperfusion of ischemic tissues. From these measurements these authors calculated a 98% reduction in nitric oxide production

A

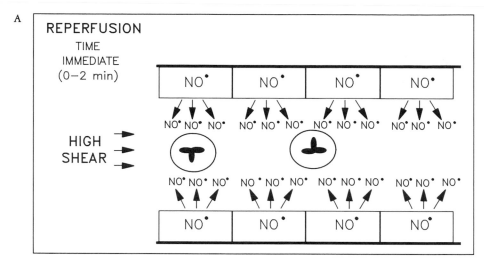

B

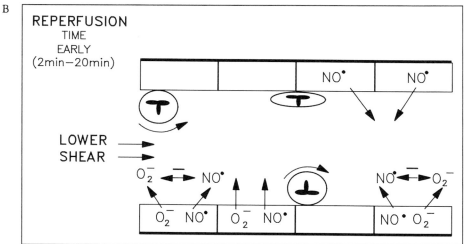

C

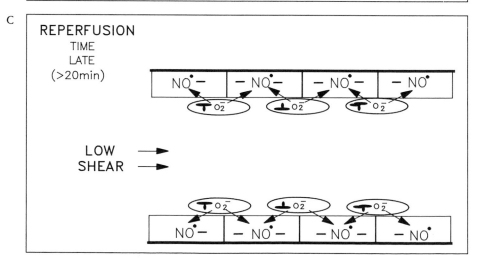

at 30 mins of reperfusion of the small intestine. Although increased superoxide levels would inactivate nitric oxide, the end products would still be nitrates and nitrites. Therefore, these data suggested that the nitric oxide synthase rather than nitric oxide per se must have been inactivated during reperfusion (Fig. 5.1C). In this particular study, L-NAME did not exacerbate I/R-induced vascular dysfunction, leukocyte adhesion and other parameters, consistent with the view that nitric oxide synthase levels were already profoundly reduced. The authors proposed that SOD may have been beneficial in previous studies because superoxide rather than inactivating nitric oxide per se was perhaps directly inactivating the nitric oxide synthase enzyme. Restoration of nitrate and nitrite levels in SOD-treated animals would provide evidence that the enzyme was indeed impaired by superoxide or superoxide-derived oxidants. This experiment has not been performed to date.

We examined nitric oxide synthase levels within the cat intestine and observed that the activity of this enzyme did indeed decrease with time of reperfusion.[14] However a significant reduction was only observed after 2 hours of I/R. Consistent with these data was our observation that L-NAME exacerbated reperfusion dysfunction at 1 but not 4 hrs of reperfusion. Simultaneously, L-arginine the precursor for nitric oxide had beneficial effects early in reperfusion but provided little protection at 4 hrs of reperfusion. These data provide enzymatic and functional evidence that nitric oxide synthase was nonfunctional but only at 4 hrs of reperfusion. A limitation of this study was the inability to determine whether the nitric oxide synthase activity was endothelial in origin or from some other cellular source. This is a critical limitation inasmuch as if the endothelial nitric oxide synthase makes up a small portion of the total nitric oxide synthase activity, early impairment of endothelial nitric oxide synthase activity may have been missed.

Fig. 5.1. (left) Panel A is a schematic of what might be occurring in the very early stages (< 2 mins) of reperfusion. A large amount of nitric oxide is released causing high shear rates and preventing leukocyte-endothelial cell interactions. However, progressively O_2^- released from the endothelium (Panel B) inhibits nitric oxide and/or nitric oxide synthase reducing shear forces and enhancing leukocyte-endothelial cell interactions. At later stages (> 20 mins) the increased leukocyte adhesion further inhibits nitric oxide and/or nitric oxide synthase via superoxide, and very little nitric oxide is now reaching target sites. Panel C shows that under these conditions shear forces are low, and leukocyte endothelial cell interactions are extensive. Whether it is the nitric oxide enzyme or nitric oxide per se that is inhibited remains an area of controversy.

One important observation of that study was the fact that the inducible nitric oxide synthase was not detected throughout the 4 hrs of reperfusion, suggesting no potential increase in nitric oxide production in this postischemic tissue.

NITRIC OXIDE DONORS REDUCE LEUKOCYTE INFLUX

The notion that nitrovasodilators may have beneficial properties in the treatment of I/R was tested many years before the discovery of endothelial-derived relaxing factor[15] or endogenous nitric oxide.[16] Initial studies to prevent I/R injury made use of nitroglycerin.[17] Since that time many different nitric oxide donors have been developed, and many of them have been tested in I/R injury, usually with some benefit.[4,13,18,19] Moreover, numerous studies have now reported that leukocyte influx could be significantly attenuated in postischemic tissues as exogenous nitric oxide levels are increased.[4,13,18,19] However the target site for the newly released nitric

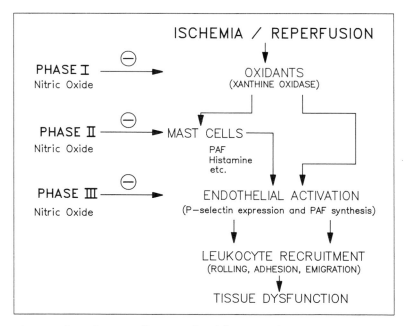

Fig. 5.2. This schematic illustrates that following ischemia/reperfusion, oxidants are produced and can activate mast cells and/or endothelium. Both mast cells and endothelium produce proinflammatory molecules that can activate CD18 on leukocytes. Additionally, P-selectin is expressed on endothelium via oxidants and mast cell-derived mediators. This leads to leukocyte rolling, adhesion and emigration and ultimately tissue injury. Nitric oxide may have inhibitory effects on oxidants (phase I), mast cells (phase II) and on endothelium (phase III). See text for details.

oxide remains a mystery. The schematic in Figure 5.2 summarizes some of the potential sites where nitric oxide may act as an antiadhesive or terminating signal for leukocyte recruitment. This section will describe in some detail the effect of nitric oxide at each of the steps in the cascade outlined in Figure 5.2.

Immediately following the reperfusion of ischemic tissue, there is an oxidative burst that may be endothelial- and likely xanthine oxidase-derived.[20] This initial increased oxidative stress may be responsible for the production of various proinflammatory mediators. The source of these mediators remains an area that requires much investigation, however there is emerging evidence that 1) endothelial cells are a source of platelet-activating factor,[21,22] and 2) the mast cell (with tremendous amounts of proinflammatory mediators released in close proximity to the microvasculature) may be a very important source of mediators (see chapter 2, Fig. 2.6, and refs. 23,24). The release of these mediators causes increased expression of endothelial-P-selectin to induce leukocyte rolling and the activation of leukocyte CD18 (β_2-integrin) which causes leukocyte adhesion.[25-27] The leukocytes emigrate out of the vasculature and ultimately cause microvascular and tissue injury. Nitric oxide may act at one or multiple sites along this cascade of events that lead to leukocyte recruitment.

NITRIC OXIDE AS AN ANTIOXIDANT

The possibility exists that nitric oxide may act as a simple antioxidant and thereby prevents reperfusion-induced leukocyte adhesion (phase 1, Fig. 5.2). Nitric oxide donors are known to rapidly react with superoxide to inactive its biologic activity.[28,29] Moreover, both nitric oxide donors and SOD have been shown to effectively inhibit leukocyte rolling[30] and adhesion in postischemic vessels.[18,20,30] The antioxidant capacity of plasma was doubled following administration of nitric oxide donors to animals at concentrations (1 mg/kg) that had antiadhesive properties,[31] however whether this relatively small increase in antioxidant capacity is physiologically important is unknown. It is possible that the nitric oxide donors dramatically increase antioxidant capacity intracellularly, however this possibility is difficult to confirm.

Administration of a superoxide-generating system caused an increase in leukocyte rolling and adhesion, and both SOD and nitric oxide donors directly inhibited leukocyte rolling and adhesion.[31] Evidence to support a correlation for nitric oxide donors as antioxidants was further provided by the fact that both SOD and nitric oxide donors demonstrated similar degrees of antiadhesion.[31] For example, both SOD and nitric oxide donors inhibited PAF-induced adhesion by approximately 50%. Moreover, oxidant-independent adhesion induced by LTB_4 resulted in no reduction in leukocyte adhesion with nitric oxide donors. Clearly, there is ample, albeit indirect evidence to support the view that nitric oxide donors may be inhibitors of leukocyte adhesion by inactivating superoxide. Finally, it should be noted that a recent report by Wink et al[32] has proposed a protective role for nitric oxide donors in cellular injury induced by hydrogen peroxide. Since catalase, the H_2O_2-detoxifying agent, can also inhibit I/R-induced leukocyte adhesion,[33] the possibility that nitric oxide donors counter H_2O_2-induced leukocyte recruitment cannot be excluded.

Although the ability of nitric oxide to react with and alter the biologic activity of superoxide is well known there is growing evidence that nitric oxide may also directly inactivate at least one of the enzymes likely to produce superoxide during I/R. The enzyme NADPH oxidase responsible for the oxidative burst that is generated by activated neutrophils is likely a very significant source of oxidants during I/R. Clancy et al[34] have recently demonstrated that nitric oxide can inhibit superoxide production from neutrophils by directly inhibiting the NADPH oxidase enzyme. Therefore, in addition to scavenging superoxide produced directly by the neutrophil, nitric oxide can prevent the synthesis of superoxide and associated oxidants including H_2O_2. The importance of this observation is that increased levels of nitric oxide could conceivably inhibit superoxide production and thereby prevent the interaction of superoxide with nitric oxide to potentially form various nitrogen intermediates including the cytotoxic peroxynitrite and hydroxyl radical (chapter 1 and ref. 35).

NITRIC OXIDE AS AN INHIBITOR OF MAST CELL REACTIVITY

There is a growing body of evidence to suggest that mast cells may be involved in the leukocyte recruitment associated with I/R (phase II, Fig. 5.2). Boros et al[36] have reported increased histamine release from the postischemic intestine and antioxidants reduced this increase in histamine release. Although the cellular source of the histamine was not identified, mast cells are a primary source of this proinflammatory mediator in the small bowel.[23] More recently, increased plasma protease II levels (an index of mast cell degranulation) have been reported during I/R.[37] Kurose et al[13] have visually assessed mast cell integrity at 30 mins reperfusion of postischemic mesentery and noted a significant increase in degranulated mast cells. The degranulated mast cells likely contribute to leukocyte recruitment inasmuch as mast cell stabilizers reduced 1) the myeloperoxidase activity in the postischemic intestine[37] and 2) the rise in leukocyte rolling and adhesion following I/R in the cat mesenteric microvasculature.[7] Although the mediator that induces most cells to degranulate during I/R remains unknown tandem administration of SOD and H_2O_2 reduced protease II levels in postischemic plasma[37] suggesting that oxidants including superoxide and catalase may activate the mast cells.

It is now well appreciated that inhibition of endogenous nitric oxide leads to mast cell degranulation and leukocyte adhesion (chapter 2 and refs. 38,39). Moreover, preventing mast cell degranulation prevents L-NAME-induced leukocyte adhesion[38] supporting the view that removal of nitric oxide causes mast cell degranulation which then recruits leukocyte adhesion. Additionally, nitric oxide donors can directly inactivate mast cell degranulation in vitro,[40,41] which can prevent mast cell-induced leukocyte adhesion to endothelial monolayers and postcapillary venules (see chapter 2, Fig. 2.5). Kurose et al[13] have recently demonstrated that the I/R-induced increase in mast cell degranulation was entirely inhibited by nitric oxide donors. In that same study, the nitric oxide donors significantly reduced leukocyte adhesion and emigration and the authors concluded that the reduction in nitric oxide production associated with I/R ultimately leads to destabilization of mast cells and the subsequent release of proadhesive molecules such as histamine, PAF and leukotrienes which could ultimately modulate the leukocyte adhesion. Addition of nitric oxide donors reversed this process. Since LTB_4-induced leukocyte adhesion has been demonstrated in I/R[42] and nitric oxide donors do not inhibit exogenous LTB_4-induced leukocyte adhesion,[31] one possible way to reconcile this difference may be that nitric oxide donors prevent the release of LTB_4 from mast cells during I/R and thereby prevent the I/R-induced adhesion. This nitric oxide-associated antiadhesive effect could occur without direct inhibition of LTB_4-induced leukocyte adhesion. Direct measurements of the release of mediators from mast cells during I/R in the presence and absence of nitric oxide donors will be necessary to confirm this hypothesis.

Nitric Oxide Inhibits Endothelial Cell Reactivity

In addition to the mast cell, it has been shown that the endothelium is capable of synthesizing various proinflammatory mediators including platelet-activating factor (Fig. 5.2). The endothelium synthesizes PAF in response to a number of stimuli including oxidants, histamine and thrombin.[21,22] PAF has been demonstrated to play an important role in the recruitment of leukocytes to postischemic tissues.[43] In fact, pretreatment of animals with PAF receptor antagonists results in almost a 90% reduction in leukocyte emigration out of the vasculature. Although emigration is entirely dependent upon the ability of the leukocyte to first firmly adhere, the PAF receptor antagonist WEB 2086 only reduced leukocyte adhesion by 40-50% (despite 90% reduction in emigration) suggesting that PAF may only partly modulate leukocyte adhesion and principally regulate emigration in postischemic tissue. Since nitric oxide donors reduce reperfusion-induced leukocyte adhesion and emigration,[13] the possibility exists that there may be an effect of nitric oxide on both PAF-induced adhesion and emigration (Phase III, Fig. 5.2).

In our laboratory, nitric oxide donors were only 50% effective in reducing PAF-induced leukocyte adhesion in postcapillary venules.[31] Therefore, it is unlikely that the partial antiadhesive effect of nitric oxide donors for PAF could explain the almost complete inhibition of adhesion and emigration of these nitric oxide releasing molecules during I/R.[18] There may exist another explanation inasmuch as nitric oxide donors have been shown to inhibit PAF synthesis by thrombin-activated endothelial cells,[44] suggesting that nitric oxide donors can directly depress the machinery responsible for PAF production. It remains to be seen whether this inhibitory effect of nitric oxide is restricted to thrombin or whether it also applies to oxidants, histamine and other mediators that are known to play a role in I/R. If nitric oxide donors universally prevent PAF synthesis in endothelium, this may be a very important event in preventing the leukocyte adhesion and emigration during I/R.

Leukocyte recruitment during I/R appears to involve increased expression of P-selectin[30,45] and activation and/or upregulation of CD18. It has been well established that P-selectin supports leukocyte rolling an event that is essential for subsequent adhesion.[46] There is some controversy as to whether nitric oxide donors can reduce P-selectin-dependent leukocyte rolling. Gauthier et al[30] have recently reported that nitric oxide donors reduce I/R-associated P-selectin expression in single rat mesenteric venules using immunohistochemistry. Associated with the decrease in P-selectin, the nitric oxide donor was also capable of completely inhibiting the increased leukocyte rolling in postischemic mesenteric venules of the rat. These data provide evidence that nitric oxide may affect P-selectin expression on endothelium. Whether this is a direct effect of nitric oxide on the endothelium or whether this is related to inhibition of mast cell degranulation which would prevent the release of potential stimulators of P-selectin such as histamine and other mediators remains unclear. Our preliminary data suggest that nitric oxide can prevent mast cell-induced, P-selectin-dependent leukocyte rolling (unpublished observations). The mast cell-induced increase in leukocyte rolling is a result of histamine,[47] yet direct exposure of postcapillary venules to the P-selectin stimulator histamine induces rolling that is not affected by nitric oxide donors. These data do not support a role for nitric oxide donors directly modulating endothelial P-selectin expression. Therefore, the possibility that nitric oxide donors affect mast cell reactivity to inhibit P-selectin expression and leukocyte rolling needs to be considered.

Interestingly, in the cat mesentery leukocyte rolling was also increased following I/R as a result of P-selectin, however nitric oxide donors did not reduce the number of rolling cells.[18] In these experiments, nitric oxide donors were added at the onset of reperfusion, and if P-selectin is expressed prior to reperfusion, the potential antirolling effect of the nitric oxide donor would have been missed. In the same series of experiments, exogenous nitric oxide dramatically reduced leukocyte adhesion during I/R. The magnitude of the reduction in fact was comparable to the inhibitory response observed with anti-CD18 antibodies. This very profound antiadhesive effect has also been reported by Gauthier et al[30] who noted that nitric oxide donors reduced leukocyte adhesion by 90%.

DOES NITRIC OXIDE PROTECT TISSUE FROM ISCHEMIA/REPERFUSION?

The last section of this chapter will consider the potential role of nitric oxide as a beneficial and/or detrimental molecule in reperfusion-induced tissue injury. I will discuss the literature as it relates to the intestine, heart and brain because of the abundance of work in these three areas. Nevertheless, there is important work on the role of nitric oxide in reperfusion of various other tissues.[19,48,49] Finally, despite the fact that up to this point the chapter has summarized literature that suggests an anti-inflammatory or antiadhesive role for nitric oxide, the beneficial role of nitric oxide in reperfusion is not clear. There are too many studies suggesting a detrimental role for nitric oxide in I/R for this to be an unambiguous issue. Therefore, I have attempted to summarize this literature as it stands to date.

INTESTINAL ISCHEMIA/REPERFUSION

Unlike the discrepant data for the role of nitric oxide in the heart or brain (next section), the role of nitric oxide in reperfusion of postischemic intestine seems to be more uniform, i.e., nitric oxide protects against reperfusion injury. The first study to demonstrate a role for nitric oxide in intestinal I/R was Aoki et al[50] who reported that a nitric oxide solution as well as acidified sodium nitrite which produces nitric oxide improved survival of animals subjected to splanchnic artery occlusion. The authors concluded that nitric oxide may prevent injury in the postischemic intestine by preventing platelet and/or neutrophil accumulation. These authors did not feel that the vasodilatory component of nitric oxide was critical to the protective effect of this molecule inasmuch as the concentrations given to animals were just above the threshold for vasodilation in cat vascular smooth muscle. The results of subsequent studies confirmed a potential beneficial role for exogenous sources of nitric oxide including SIN-1, CAS 754 and nitroprusside in I/R injury.[14,51] Moreover, the data from these experiments also suggested that alterations in blood flow were an unlikely event in the cytoprotective effect of nitric oxide. SIN-1, CAS 754 and nitroprusside all had protective effects on the mucosa despite reperfusion-induced decreases in intestinal blood flow in these experimental groups that did not differ from the untreated group.[51] It should be noted, however, that SIN-1, CAS 754 and nitroprusside decreased systemic blood pressure, and intestinal vascular resistance did not rise in these groups as it did in the experimental group. Nevertheless, L-arginine also protected against reperfusion injury of the mucosa without affecting intestinal hemodynamic parameters. By contrast, improving blood flow did not always improve postischemic tissue injury. For example, 8-bromo-cyclic GMP prevented the reperfusion-induced drop in intestinal blood flow and the increase in vascular resistance, yet the mucosal dysfunction was not reduced.[51] These data strongly suggest that the beneficial effect of nitric oxide donors as well as L-arginine in the postischemic intestine is not a result of improved perfusion.

Recent evidence would suggest that nitric oxide may affect both leukocyte recruitment and platelet aggregation and thereby reduce intestinal injury. In the rat mesentery Kurose et al[13] tested three different nitric oxide donors with very similar results; all of the donors reduced leukocyte adhesion by approximately 50-60%. The adhesion correlated very closely with microvascular dysfunction (r = 0.8, p < 0.05) suggesting that the ability of the nitric oxide donors to blunt leukocyte adhesion was related to their ability to prevent the increased microvascular permeability. In the same study, exposure of the intestine to I/R resulted in the appearance of large platelet-leukocyte aggregates that coursed through the vessel lumen without necessarily making adhesive contact with endothelium. These aggregates have previously been reported in mesenteric vessels exposed to nitric oxide synthase inhibitors,[52] further supporting the view that nitric oxide inhibition and I/R have many sequelae in common. Addition of nitric oxide donors to the postischemic vessels inhibited the formation of these aggregates by approximately 90%.[13] The potential inflammatory and tissue injuring role of these aggregates is presently unknown, but their formation appears to be a direct result of a lack of nitric oxide within the vasculature. Zhang et al[53] did not observe an antiaggregatory effect of nitroprusside on platelets in a cerebral ischemia model. However in that study, the platelet aggregation was measured ex vivo and not directly in the animals undergoing I/R.

Inhibition of nitric oxide synthesis has been reported to exacerbate damage to the intestine during partial[54] or complete[51] I/R of the intestine. These data albeit indirect support the view that nitric oxide may be an endogenous molecule with beneficial properties during reperfusion. D-NAME, the biologically inactive enantiomer, did not enhance intestinal dysfunction, whereas L-arginine, the substrate for nitric oxide, reversed the effects of L-NAME. These data led us to conclude that the presence of endogenous nitric oxide may have beneficial properties during the inflammatory process, however the mechanism remained unclear.

Addition of nitric oxide synthesis inhibitor L-NAME at 1 hr of reperfusion enhanced mucosal dysfunction very significantly.[54] These data suggest that the available endogenous nitric oxide produced at 1 hr of reperfusion was essential for preserving the integrity of the mucosal barrier, and in its absence the mucosa is further injured.[54] In this study, the enhanced injury associated with L-NAME could be prevented with L-arginine, nitric oxide donors, as well as 8-bromo-cyclic GMP. The latter suggests that cyclic GMP may be involved in maintaining a tight mucosal barrier. It is well known that nitric oxide activates the soluble or cytosolic form of guanylate cyclase which increases the conversion of GTP to cyclic GMP,[55] a second messenger found in all cells. Moreover, cyclic GMP reacts with cyclic GMP-dependent protein kinase and causes a cascade of changes in protein phosphorylation, resulting in dephosphorylation of myosin light chain which leads to cellular relaxation.[55-57] The absence of nitric oxide (L-NAME-induced) might reduce cyclic GMP levels and cause epithelial cell contraction which would produce larger interepithelial junctions, an event that results in a leakier mucosal barrier.[57]

As already mentioned however, the permeable analog of cyclic GMP, 8-bromo-cyclic GMP, did not reduce the reperfusion-induced leaky mucosal barrier despite complete reversal of L-NAME-induced increases in permeability.[54] Clearly, improving cGMP levels does not protect the reperfused mucosa, suggesting that the protective effect of nitric oxide donors in reperfusion-induced rise in epithelial permeability is unlikely to be associated with improved cGMP levels. Although many nitric oxide-dependent mechanisms function via cGMP, inactivation of superoxide for example does not require the second messenger. Further elucidation of the nitric oxide-dependent, cGMP-independent protective mechanism is required.

CEREBRAL ISCHEMIA/REPERFUSION

Unlike the intestine, nitric oxide donors may be critical in maintaining blood flow to the brain following reperfusion. Zhang et al[53] reported that nitric oxide donors enhance blood flow to the ischemic territory and thereby attenuate focal ischemic damage. These studies have been confirmed by a number of investigators who have shown that nitric oxide production increased blood flow to marginally ischemic tissues.[58,59] In addition, Kuluz et al[60] have demonstrated that inhibition of nitric oxide synthesis increases motor deficits, infarct size and decreases body weight. Others have reported that inhibition of endogenous nitric oxide either did not affect infarct size or reduced infarct size.[61,62] Such dichotomous findings between studies are intriguing and differences such as dissimilar strains of rats or differences in time of ischemia seem unlikely explanations for such diverse results. Moreover, dichotomous results have been obtained in short experiments as well as in chronic recovery experiments that have lasted days. Therefore, it is difficult to invoke time as a variable in these studies.

One pattern is emerging that may provide some insight into the controversial results. Ameliorative effects are observed when a low dose of nitric oxide synthase inhibitors are administered, whereas high doses of nitric oxide synthase inhibitors have not been protective and often have shown deleterious effects. A dose response study revealed a very narrow window in which nitric oxide synthase inhibitors may protect during I/R; 3 mg/kg but not 1 or 10 mg/kg provided a statistically significant neuroprotective effect in the postischemic gerbil brain.[63] These authors raised the possibility that reducing the overstimulation of nitric oxide synthase from neural cells with low concentrations of N^w-nitro-L-arginine may be beneficial during reperfusion of the cerebral vasculature whereas high concentrations of N^w-nitro-L-arginine may prevent physiologic vascular tone and thereby exacerbate injury. This hypothesis was recently tested and to some extent confirmed. Huang et al[64] used mutant mice that did not express the gene for the neuronal isoform of nitric oxide synthase and documented a reduction in infarct areas and neurological deficits but no difference in regional blood flow following middle cerebral artery occlusion. These data clearly demonstrate that neuronal nitric oxide production seems to exacerbate ischemic injury in the brain. However when the mutant animals received a nitric oxide synthase inhibitor, infarct size was greatly enhanced suggesting that constitutive nonneuronal nitric oxide synthase is very important in maintaining perfusion and other homeostatic properties in the vasculature. Although nitric oxide donors were not tested in this study, it would be intriguing to determine whether these agents harmed the cerebral microcirculation. The role of nitric oxide in brain I/R will not be resolved with presently available nitric oxide synthase inhibitors, but new more-selective inhibitors as well as the production of mutant mice lacking each of the nitric oxide synthases may go a long way in helping us understand the role of nitric oxide in cerebral I/R.

MYOCARDIAL ISCHEMIA/REPERFUSION

The role of nitric oxide in I/R of the heart is also somewhat controversial. Numerous investigators have reported a protective effect of nitric oxide donors following I/R of the heart. Lefer et al[65] demonstrated that the nitric oxide donor, SPM-5185, restored regional function of the postischemic heart, reduced myocardial necrosis and significantly reduced myeloperoxidase activity in both the ischemic zone and the necrotic zone. Although there was some improvement in postischemic blood flow in the SPM-5185 group which could potentially explain some of the protective effect of the nitric oxide donor, others have used nitric oxide or nitric oxide donors at subvasodilator concentrations and also observed a significant reduction of myocardial necrosis during reperfusion.[66] Therefore the vasodilatory effects of nitric oxide donors are unlikely to explain the cardioprotective effects. Many of these authors have concluded that it is the reduction in leukocyte infiltration that protects the heart from I/R injury.[65,66] To our knowledge nitric oxide donors have not been observed to exacerbate myocardial reperfusion injury.

The nitric oxide synthesis inhibition data are not as clear. A number of laboratories have reported that nitric oxide synthesis inhibition enhances myocardial reperfusion injury. Fung et al[67] observed enhanced necrosis and an increased pressure-rate index following inhibition of nitric oxide synthesis during reperfusion. Ma et al[12] observed increased leukocyte adhesion in postischemic coronary vessels following nitric oxide synthesis inhibition suggesting a potentially important role for endogenous nitric oxide as a cytoprotective and antiadhesive molecule.

By contrast Matheis et al[68] reported that endogenous nitric oxide may be responsible for myocardial reoxygenation injury in the hypoxic piglet on cardiopulmonary bypass. These authors demonstrated that the nitric oxide synthase inhibitor, L-NAME, afforded nearly complete protection against myocardial reoxygenation injury, an effect entirely nullified by adding excess L-arginine. Additionally, these authors demonstrated that systemic venous and coronary sinus blood content of nitric oxide (nitrite) increased substantially during reoxygenation, consistent with the view that increased nitric oxide production could conceivably damage myocardial tissue. These authors demonstrated an increase in conjugated diene content in reoxygenated endocardium. L-NAME as well as various antioxi-

dants (catalase and mercaotopropionyl glycine) eliminated the increase
in this marker of lipid peroxidation. Matheis et al[68] concluded that
the L-arginine/nitric oxide pathway was contributing to the myocardial reoxygenation injury by generating nitrogen intermediates including possibly peroxynitrite (see chapter 1). Unfortunately, inhibition of nitric oxide synthase with an L-arginine analog resulted in
severe systemic vasoconstriction and pancreatic damage. Therefore,
these investigators took a different approach in a subsequent study
to minimizing myocardial injury following reoxygenation without
injuring other organs. They reported[69] that whereas rapid
reoxygenation caused depression in myocardial function, gradual
reoxygenation was without pathology. Moreover, rapid
reoxygenation caused a 10-fold rise in cardiac nitric oxide production whereas gradual reoxygenation was unremarkable as an enhancer
of nitric oxide levels. The authors concluded that reoxygenation
injury was related to increased nitric oxide production and that
controlled cardiac reoxygenation may limit the increase in nitric oxide production and thereby improve myocardial status postopera-
tively. The authors proposed a nitric oxide paradox whereby nitric
oxide is essential as a regulator of several physiologic functions but
becomes deleterious when produced in high concentrations.

It is interesting to note that in complete contrast, cardiac preservation is enhanced in a heterotopic rat transplant model by
supplementing the nitric oxide pathway. In this study, Pinsky et
al[4] demonstrated that only 17% of hearts survived a 12 hrs storage period (followed by reperfusion) in lactated Ringer's solution
whereas addition of a nitric oxide donor to the solution improved
survival to 92%. Addition of nitroglycerin to the clinical standard
University of Wisconsin solution improved survival from 35% to
100%. Nitroprusside, L-arginine and 8-bromo-cGMP all dramatically improved survival whereas two structurally distinct nitric oxide synthase inhibitors guaranteed failure following transplantation
(0% survival). These investigators demonstrated that nitric oxide
levels were greatly depressed in rat hearts subjected to preserva-
tion/heterotopic transplantation using a specific nitric oxide-sens-
ing electrode. Superoxide dismutase abrogated the impairment in nitric oxide levels suggesting that the nitric oxide was being quenched
by superoxide. Indeed measurement of nitric oxide synthase revealed
similar or even enhanced biologic activity of the enzyme in transplanted hearts negating the likelihood of impaired nitric oxide synthase

activity in the transplanted organs. The addition of nitric oxide donors to the preservation fluid produced two interesting observations; 1) blood flow was improved almost 4-fold relative to untreated counterparts, and 2) both histology and the myeloperoxidase assay revealed a 10-fold increase in leukocyte accumulation in untreated versus nitroglycerin-treated transplanted hearts. The authors concluded that supplementation of nitric oxide to these organs improves blood flow and decreases leukostasis thereby improving graft survival. This study has now been confirmed in orthotopic baboon cardiac transplantation[70] wherein supplementation of a storage solution with nitroglycerin and a cAMP analog preserved hearts successfully for an unprecedented 24 hrs. This may provide a new approach to enhancing cardiac preservation.

The obvious differences between the two series of studies described above[4,69] are striking but an obvious explanation for the discrepancy remains unavailable. One could argue that in the cardiobypass work too much nitric oxide was produced, whereas in the transplantation work insufficient nitric oxide was produced. However, eliminating nitric oxide in cardiobypass or providing excess nitric oxide in transplantation results in beneficial effects making it difficult to accept the thesis that too much or too little nitric oxide is detrimental. Both experiments likely have a very similar reperfusion profile associated with neutrophil influx, oxidative stress and microvascular dysfunction, yet nitric oxide has diametrically opposed effects on these two models. One possible explanation is that leukocytes do not contribute to or even infiltrate the myocardium in the cardiobypass study, thereby making it a different type of injury process from the transplant model. The challenge in the next few years will be to resolve this obviously real discrepancy.

References

1. Sternbergh WC, Makhoul RG, Adelman B. Nitric oxide-mediated, endothelium-dependent vasodilation is selectively attenuated in the postischemic extremity. Surgery 1993; 114:960-967.
2. Lefer AM, Tsao PS, Lefer DJ et al. Role of endothelial dysfunction in the pathogenesis of reperfusion injury after myocardial ischemia. FASEB J 1991; 5:2029-2034.
3. Tsao PS, Aoki N, Lefer DJ et al. Time course of endothelial dysfunction and myocardial injury during myocardial ischemia and reperfusion in the cat. Circulation 1990; 82:1402-1412.
4. Pinsky DJ, Oz MC, Koga S et al. Cardiac preservation is enhanced

in a heterotopic rat transplant model by supplementing the nitric oxide pathway. J Clin Invest 1994; 93:2291-2297.

5. Ma X-L, Johnson G, Lefer AM. Mechanisms of inhibition of nitric oxide production in a murine model of splanchnic artery occlusion shock. Arch Int Pharmacodyn 1991; 311:89-103.

6. Mayhan WG, Amundesen SM, Faraci FM et al. Responses of cerebral arteries after ischemia and reperfusion in cats. Am J Physiol 1988; 255:H879-H884.

7. Kanwar S, Kubes P. Ischemia/reperfusion-induced granulocyte influx is a multistep process mediated by mast cells. Microcirc 1994; 1(3):175-182.

8. Lefer AM, Aoki N. Leukocyte-dependent and leukocyte-independent mechanisms of impairment of endothelium-mediated vasodilation. Blood Vessels 1990; 27:162-168.

9. Ma X-L, Tsao PS, Viehman GE et al. Neutrophil-mediated vasoconstriction and endothelial dysfunction in low-flow perfusion-reperfused cat coronary artery. Circ Res 1991; 69:95-106.

10. Albertine KH, Weyrich AS, Ma X et al. Quantification of neutrophil migration following myocardial ischemia and reperfusion in cats and dogs. J Leukoc Biol 1994; 55:557-566.

11. Kubes P, Suzuki M, Granger DN. Nitric oxide: An endogenous modulator of leukocyte adhesion. Proc Natl Acad Sci USA 1991; 88:4651-4655.

12. Ma X-L, Weyrich AS, Lefer DJ et al. Diminished basal nitric oxide release after myocardial ischemia and reperfusion promotes neutrophil adherence to coronary endothelium. Circ Res 1993; 72:403-412.

13. Kurose I, Wolf R, Grisham MB et al. Modulation of ischemia/reperfusion-induced microvascular dysfunction by nitric oxide. Circ Res 1994; 74:376-382.

14. Kanwar S, Tepperman BL, Payne D et al. Time course of nitric oxide production and epithelial dysfunction during ischemia/reperfusion of the feline small intestine. Circ Shock 1994; 42:135-140.

15. Furchgott RF, Zawadzki JV. The obligatory role of endothelial cells in the relaxation of arterial smooth muscle by acetylcholine. Nature 1980; 288:373-376.

16. Ignarro LJ, Buga GM, Wood KS et al. Endothelium-derived relaxing factor produced and released from artery and vein is nitric oxide. Proc Natl Acad Sci USA 1987; 84:9265-9269.

17. Amstrong PW, Walker DC, Burton JR et al. Vasodilator therapy in acute myocardial infarction: A comparison of sodium nitroprusside and nitroglycerin. Circulation 1975; 52:1118-1122.

18. Kubes P, Kurose I, Granger DN. NO donors prevent integrin-induced leukocyte adhesion, but not P-selectin-dependent rolling in postischemic venules. Am J Physiol 1994; 267:H931-H937.

19. Abdih H, Kelly CJ, Bouchier-Hayes D et al. Nitric oxide (endothelium-derived relaxing factor) attenuates revascularization-induced lung injury. J Surg Res 1994; 57:39-43.
20. Granger DN. Role of xanthine oxidase and granulocytes in ischemia-reperfusion injury. Am J Physiol 1988; 255:H1269-H1275.
21. Lorant DE, Patel KD, McIntyre TM et al. Coexpression of GMP-140 and PAF by endothelium stimulated by histamine or thrombin: a juxtacrine system for adhesion and activation of neutrophils. J Cell Biol 1991; 115:223-234.
22. Lewis MS, Whatley RE, Cain P et al. Hydrogen peroxide stimulates the synthesis of platelet activating factor by endothelium and induces endothelial cell-dependent neutrophil adhesion. J Clin Invest 1988; 82:2045-2055.
23. Crowe SE, Perdue MH. Gastrointestinal food hypersensitivity: Basic mechanisms of pathophysiology. Gastroenterology 1992; 103:1075-1095.
24. Galli SJ. New concepts about the mast cell. N Engl J Med 1993; 328:257-265.
25. Tonnesen MG. Neutrophils-endothelial cell interactions: mechanisms of neutrophil adherence to vascular endothelium. J Invest Dermatol 1989; 93:53s-58s.
26. Bevilacqua MP, Nelson RM. Selectins. J Clin Invest 1993; 91:379-387.
27. Granger DN, Kubes P. The microcirculation and inflammation: modulation of leukocyte-endothelial cell adhesion. J Leukocyte Biol 1994; 55:662-675.
28. Rubanyi GM, Vanhoutte PM. Superoxide anions and hyperoxia inactivate endothelium-derived relaxing factor. Am J Physiol 1986; 250:H822-H827.
29. Rubanyi GM, Ho EH, Cantor EH et al. Cytoprotective function of nitric oxide: inactivation of superoxide radicals produced by human leukocytes. Biochem Biophys Res Commun 1991; 181:1392-1397.
30. Gauthier TW, Davenpeck KL, Lefer AM. Nitric oxide attenuates leukocyte-endothelial interaction via P-selectin in splanchnic ischemia-reperfusion. Am J Physiol 1994; 267:G562-G568.
31. Gaboury J, Woodman RC, Granger DN et al. Nitric oxide prevents leukocyte adherence: role of superoxide. Am J Physiol 1993; 265:H862-H867.
32. Wink DA, Hanbauer I, Krishna MC et al. Nitric oxide protects against cellular damage and cytotoxicity from reactive oxygen species. Proc Natl Acad Sci USA 1993; 90:9813-9817.
33. Suzuki M, Grisham MB, Granger DN. Leukocyte-endothelial cell adhesive interactions: Role of xanthine oxidase-derived oxidants. J Leukocyte Biol 1991; 50:488-494.
34. Clancy RM, Leszczynska-Piziak J, Abramson SB. Nitric oxide, and endothelial cell relaxation factor, inhibits neutrophil superoxide production via a direct action on the NADPH oxidase. J Clin In-

vest 1992; 90:1116-1121.

35. Beckman JS, Beckman TW, Chen J et al. Apparent hydroxyl radical production by peroxynitrite: Implications for endothelial injury from nitric oxide and superoxide. Proc Natl Acad Sci USA 1990; 87:1620-1624.

36. Boros M, Kaszaki J, Nagy S. Histamine release during intestinal ischemia-reperfusion: Role of iron ions and hydrogen peroxide. Circ Shock 1991; 35:174-180.

37. Kanwar S, Kubes P. Mast cells contribute to ischemia/ reperfusion-induced granulocyte infiltration and intestinal dysfunction. Am J Physiol 1994; 267:G316-G321.

38. Kubes P, Kanwar S, Niu X-F et al. Nitric oxide synthesis inhibition induces leukocyte adhesion via superoxide and mast cells. FASEB J 1993; 7:1293-1299.

39. Kanwar S, Wallace JL, Befus D et al. Nitric oxide synthesis inhibition increases epithelial permeability via mast cells. Am J Physiol 1994; 266:G222-G229.

40. Salvemini D, Masini E, Pistelli A et al. Nitric oxide: A regulatory mediator of mast cell reactivity. J Cardiovasc Pharmacol 1991; 17 (Suppl 3):S258-S264.

41. Hogaboam CM, Bissonnette EY, Befus AD et al. Modulation of rat mast cell reactivity by IL-1 beta. Divergent effects on nitric oxide and platelet-activating factor release. J Immunol 1993; 151:3767-3774.

42. Zimmerman BJ, Guillory DJ, Grisham MB et al. Role of leukotriene B_4 in granulocyte infiltration into the postischemic feline intestine. Gastroenterology 1993; 99:1-6.

43. Kubes P, Ibbotson G, Russell JM et al. Role of platelet-activating factor in ischemia/reperfusion-induced leukocyte adherence. Am J Physiol 1990; 259:G300-G305.

44. Heller R, Bussolino F, Ghigo D et al. Nitrovasodilators inhibit thrombin-induced platelet-activating factor synthesis in human endothelial cells. Biochem Pharmacol 1992; 44:223-229.

45. Davenpeck KL, Gauthier TW, Albertine KH et al. Role of P-selectin in microvascular leukocyte-endothelial interaction in splanchnic ischemia-reperfusion. Am J Physiol 1994; 267:H622-H630.

46. Lawrence MB, Springer TA. Leukocytes roll on a selectin at physiologic flow rates: distinction from and prerequisite for adhesion through integrins. Cell 1991; 65:859-873.

47. Gaboury JP, Johnston B, Niu X-F et al. Mechanisms underlying acute mast cell-induced leukocyte rolling and adhesion in vivo. J Immunol 1995; 154:804-813.

48. Jaeschke H, Schini VB, Farhood A. Role of nitric oxide in the oxidant stress during ischemia/reperfusion injury in the liver. Life Sciences 1992; 50:1797-1804.

49. Andrews FJ, Malcontenti-Wilson C, O'Brien PE. Protection against gastric ischemia-reperfusion injury by nitric oxide generators. Dig Dis Sci 1994; 39:366-373.

50. Aoki N, Johnson III G, Lefer AM. Beneficial effects of two forms of NO administration in feline splanchnic artery occlusion shock. Am J Physiol 1990; 258:G275-G281.

51. Payne D, Kubes P. Nitric oxide donors reduce the rise in reperfusion-induced intestinal mucosal permeability. Am J Physiol 1993; 265:G189-G195.

52. Kurose I, Kubes P, Wolf R et al. Inhibition of nitric oxide production: Mechanisms of vascular albumin leakage. Circ Res 1993; 73:164-171.

53. Zhang F, White JG, Iadecola C. Nitric oxide donors increase blood flow and reduce brain damage in focal ischemia: evidence that nitric oxide is beneficial in the early stages of cerebral ischemia. J Cereb Blood Flow Metab 1994; 14:217-226.

54. Kubes P. Ischemia/reperfusion in the feline small intestine: a role for nitric oxide. Am J Physiol 1993; 264:G143-G149.

55. Ignarro LJ, Harbison RG, Wood KS et al. Activation of purified soluble guanylate cyclase by endothelium-derived relaxing factor from intrapulmonary artery and vein: stimulation by acetylcholine, bradykinin and arachidonic acid. J Pharm Exp Ther 1986; 237:893-237.

56. Wysolmerski RB, Lagunoff D. Involvement of light-chain kinase in endothelial cell retraction. Proc Natl Acad Sci USA 1990; 87:16-20.

57. Madara JL, Moore R, Carlson S. Alterations in intestinal tight junction structure and permeability by cytoskeleton contraction. Am J Physiol 1987; 253:C854-C861.

58. Yamamoto S, Golanov EV, Berger SB et al. Inhibition of nitric oxide synthesis increases focal ischemic infarction in rat. J Cereb Blood Flow Metab 1992; 12:717-726.

59. Morikawa E, Huang Z, Moskowitz MA. L-Arginine decreases infarct size caused by middle cerebral arterial occlusion in SHR. Am J Physiol 1992; 263:H1632-H1635.

60. Kuluz JW, Prado RJ, Dietrich D et al. The effect of nitric oxide synthase inhibition on infarct volume after reversible focal cerebral ischemia in conscious rats. Stroke 1993; 24:2023-2029.

61. Nowicki JP, Duval D, Poignet H et al. Nitric oxide mediates neuronal death after focal cerebral ischemia in the mouse. Eur J Pharmacol 1991; 204:339-340.

62. Dawson DA, Kusumoto K, Graham DI et al. Inhibition of nitric oxide synthesis does not reduce infarct volume in a rat model of focal cerebral ischaemia. Neurosci Lett 1992; 142:151-154.

63. Nagafuji T, Sugiyama M, Matsui T et al. A narrow therapeutic window of a nitric oxide synthase inhibitor against transient ischemic brain injury. Eur J Pharmacol 1993; 248:325-328.

64. Huang Z, Huang PL, Panahian N et al. Effects of cerebral ischemia in mice deficient in neuronal nitric oxide synthase. Science 1994; 265:1883-1885.

65. Lefer DJ, Nakanishi K, Johnston WE et al. Antineutrophil and myocardial protecting actions of a novel nitric oxide donor after acute myocardial ischemia and reperfusion in dogs. Circulation 1993; 88, no. 5 part 1:2337-2350.

66. Siegfried MR, Erhardt J, Rider T et al. Cardioprotection and attenuation of endothelial dysfunction by organic nitric oxide donors in myocardial ischemia-reperfusion. J Pharm Exp Ther 1992; 260:668-675.

67. Fung KP, Wu TW, Zeng LH et al. The opposing effects of an inhibitor of nitric oxide synthesis and of a donor of nitric oxide in rabbits undergoing myocardial ischemia reperfusion. Life Sciences 1994; 54:PL491-PL496.

68. Matheis G, Sherman MP, Buckberg GD et al. Role of L-arginine-nitric oxide pathway in myocardial reoxygenation injury. Am J Physiol 1992; 262:H616-H620.

69. Morita K, Ihnken K, Buckberg GD et al. Role of controlled cardiac reoxygenation in reducing nitric oxide production and cardiac oxidant damage in cyanotic infantile hearts. J Clin Invest 1994; 93:2658-2666.

70. Oz MC, Pinsky DJ, Koga S et al. Novel preservation solution permits 24-hour preservation in rat and baboon cardiac transplant models. Circulation 1993; 88:291-297.

71. Weyrich AS, Ma X-L, Lefer DJ et al. In vivo neutralization of P-selectin protects feline heart and endothelium in myocardial ischemia and reperfusion injury. J Clin Invest 1993; 91: 2620-2629.

72. Davenpeck KL, Gauthier TW, Lefer AM. Inhibition of endothelial-derived nitric oxide promotes P-selectin expression and actions in the rat microcirculation. Gastroenterology 1994; 107:1050-1058.

73. Granger DN, Benoit JN, Suzuki M et al. Leukocyte adherence to venular endothelium during ischemia-reperfusion. Am J Physiol 1989; 257:G683-G688.

74. Arndt H, Russell JM, Kurose I et al. Mediators of leukocyte adhesion in rat mesenteric venules elicited by inhibition of nitric oxide synthesis. Gastroenterology 1993; 105:675-680.

75. Welbourn R, Goldman G, Kobzik L et al. Neutrophil adherence receptors (CD18) in ischemia. Dissociation between quantitative cell surface expression and disapedesis mediated by leukotriene $B_4{}^1$. J Immunol 1991; 145:1906-1911.

76. Nilsson Ua, Lundgren O, Haglind E, Bylund-Fellienius A-C. Radical production during intestinal ischemia and reperfusion in vivo in the cat--an ESR study. In: Simic M, ed. Proceedings of 4th International Congress on Oxygen Radicals. 1987:150-152.

77. Roldan EJA, Pinus CR, Turrens JF et al. Chemiluminescence of ischaemic and reperfused intestine in vivo. GUT 1989; 30:184-187.

78. Kurose I, Wolf R, Grisham MB et al. Microvascular responses to inhibition of nitric oxide production: role of active oxidants. Circ Res 1995; 76:30-39.

79. Suematsu M, Tamatani T, Delano FA et al. Microvascular oxidative stress preceding leukocyte activation elicited by in vivo nitric oxide suppression. Am J Physiol 1994; 266:H2410-H2415.

80. Yokota K, Takishima T, Sato K et al. Comparative studies of FK506 and cyclosporine in canine orthotopic hepatic allograft survival. Transplant Proc 1989; 21:1066-1068.

NITRIC OXIDE AS A MODULATOR OF LEUKOCYTE FUNCTION IN ATHEROSCLEROSIS

Dwight D. Henninger, Lianxi Liao,
Iwao Kurose and D. Neil Granger

INTRODUCTION

Atherosclerosis, which represents the principal cause of death in the USA and Europe, is a complex disease process involving interactions between various circulating blood cells, plasma components (e.g., lipoproteins), and the blood vessel wall. Although several hypotheses have been offered to explain the process of atherogenesis, the "response to injury hypothesis," which proposes that endothelial cell injury is the initiating event in atherosclerosis, is the most widely accepted because it serves to unify the multiple causative factors and their potential role in the disease process.[1] According to this hypothesis the initiating endothelial cell dysfunction and injury may result from a mechanical, immunologic, and/or biochemical insult.[1] In the presence of a hyperlipidemic environment, this insult elicits a chronic inflammatory condition characterized by an inflammatory cell infiltrate, impaired endothelium-dependent vascular relaxation,

Nitric Oxide: A Modulator of Cell-Cell Interactions in the Microcirculation, edited by
Paul Kubes. © 1995 R.G. Landes Company.

and vascular smooth muscle proliferation, with possible end-stage occlusion of the vessel.

An important manifestation of the endothelial cell dysfunction that occurs early in the pathogenesis of atherosclerosis is the loss of endothelial cell-derived relaxing factor (EDRF), which is now recognized to be nitric oxide (NO). NO is a potent vasodilator that also appears to contribute to the physiologic modulation of leukocyte-endothelial cell adhesion, platelet adherence and aggregation, and vascular smooth muscle proliferation.[2] There is a growing body of evidence that implicates the loss of endothelial cell-derived NO as a key event that contributes to the inflammatory response and vascular dysfunction associated with inflammation. In this chapter, we summarize the available data that invokes a role for NO in the exaggerated leukocyte-endothelial cell adhesion and other abnormal vascular responses observed in hypercholesterolemic (HCh) and/or atherosclerotic animals.

IMPAIRED ENDOTHELIUM-DEPENDENT RESPONSES IN HYPERCHOLESTEROLEMIA AND ATHEROSCLEROSIS

The release of NO or a closely related compound from vascular endothelium is generally considered to play a major role in the regulation of vasomotor tone. The development of hypertension, altered tissue blood flow, and enhanced propensity for vasoconstriction associated with some disease states has been attributed to a defective metabolism (reduced production, increased degradation) of endothelial cell-derived NO. Studies in man and different experimental animals have revealed an association between atherosclerosis (also hypercholesterolemia or hyperlipoproteinemia) and impaired endothelium-dependent vasorelaxation. For example, aortae isolated from cholesterol-fed rabbits with atherosclerosis relax less to the endothelium-dependent vasodilator acetylcholine than aortae derived from their normocholesterolemic counterparts.[3] These differences cannot be attributed to structural alterations in the vasculature that normally accompanies atherosclerosis inasmuch as the impaired vasodilatory capacity is manifested even before the development of atherosclerotic lesions. Furthermore, the altered vascular responsiveness associated with hypercholesterolemia is also observed in the microvasculature, i.e., there is a defect in endothelium-dependent vascular relaxation that can be demonstrated in arterioles of 25 μm diameter.[4] Since the microvasculature responds normally to endothelium-independent dilators (adenosine, nitroprusside), it would appear that hypercholesterolemia affects the metabolism (production and/or degradation) of NO by endothelial cells, rather than affecting the sensitivity of vascular smooth muscle.

MECHANISMS RESPONSIBLE
FOR DEFECTIVE ENDOTHELIAL CELL GENERATION
OF NITRIC OXIDE IN ATHEROSCLEROSIS

The biochemical and molecular basis for the defective NO generation and corresponding impairment of endothelium-dependent vasodilation in atherosclerosis remains undefined. However, the mechanisms that have been proposed to date place emphasis on either a defective production of NO from L-arginine or an enhanced inactivation of NO by superoxide or other oxygen-centered radicals. The concept that deficiencies in substrate (L-arginine) and/or cofactors (tetrahydrobiopterin) for endothelial cell nitric oxide synthase (NOS) may account for the diminished production of NO in atherosclerosis has received considerable attention. It has been repeatedly shown that administration of exogenous L-arginine improves the responses of large vessels and the microvasculature to endothelium-dependent dilators and attenuates the platelet hyper-reactivity normally observed in both humans and animals with hypercholesterolemia.[5] Furthermore, chronic administration of an L-arginine supplemented diet to HCh rabbits limits the development of aortic lesions,[6] while chronic inhibition of NO production (with L-NAME) accelerates neointimal formation and further impairs endothelial cell function in HCh animals.[7] L-arginine supplementation in normocholesterolemic experimental animals and humans rarely leads to enhanced vascular and platelet functions.

While reduced plasma levels of L-arginine have been noted in HCh patients, it is generally argued that the decline in plasma L-arginine levels is not sufficient to be rate-limiting for NO production.[4] It is conceivable however that L-arginine levels within endothelial cells are reduced as a consequence of impaired amino acid transport induced by hypercholesterolemia. Alternatively, hypercholesterolemia may increase the K_m of NOS for L-arginine or it may limit the availability of obligatory cofactors such as tetrahydrobiopterin (BH_4). BH_4 is constitutively secreted in large quantities by human and murine endothelial cells.[8] Since intracellular availability of BH_4 is rate-limiting for NO synthesis in endothelial cells, an imbalance between the rates of synthesis of the cofactor and its export from the cell (secretion) could account for the decline in NO generation associated with hypercholesterolemia.

The possibility that NOS activity is inhibited in hypercholesterolemia and atherosclerosis has some merit. The fact that the defective endothelium-dependent vascular responses in atherosclerosis are observed in segments of the circulation where lipid deposition (and atherosclerotic lesions) does not occur, i.e., the microcirculation, has led to the proposal that hypercholesterolemia exerts it effects via circulating factors such as oxidized low density lipoproteins (oxLDL)[9] and L-dimethylarginine (L-DMA).[10] Indeed, exposure of the microvasculature to either of these endogenous agents (to mimic levels measured in hypercholesterolemia or atherosclerosis) elicits many of the same deleterious responses observed in hypercholesterolemia. The ability of oxLDL to induce these microvascular alterations may be related to the observation that oxLDL appears to destroy NO in a manner similar to superoxide[11] and also inhibits both the constitutive[12] and inducible[13] forms of NOS. L-DMA is a naturally occurring NOS inhibitor that is found to be significantly elevated in the plasma of rabbits with hypercholesterolemia, compared with their normocholesterolemic counterparts.[10] The actions of oxLDL and L-DMA on leukocyte-endothelial cell adhesion and other microvascular functions are described below.

REACTIVE OXYGEN METABOLITES IN HYPERCHOLESTEROLEMIA AND ATHEROSCLEROSIS

The results of several in vivo studies support the view that the production of reactive oxygen metabolites (ROM) is enhanced during the pathogenesis of atherosclerosis. It has been shown that HCh rats are more prone to an oxidant stress than their normocholesterolemic counterparts, as reflected by lower serum levels of vitamin E, elevated levels of thiobarbituric acid reactive products (TBARs), and increased exhaled pentane, with the latter two indices reflecting elevated lipid peroxidation.[14] ROM have also been implicated in the pathogenesis of HCh-induced atherosclerosis in rabbits. In this animal model, there is an elevated concentration of malondialdehyde (an indicator of lipid peroxidation), increased superoxide production by circulating neutrophils, and reductions in the activities of superoxide dismutase (SOD) and glutathione peroxidase in blood.[15] All of these effects are restored to normal, without a change in plasma cholesterol, when the HCh rabbits are also placed on a vitamin E-supplemented diet.

A role for superoxide in the pathobiology of atherosclerosis is supported by the observation that SOD significantly restores the defective endothelium-dependent vascular relaxation that is observed in HCh rabbits.[16] This possibility is supported by evidence that aortic endothelial cells from HCh rabbits produce significantly greater quantities of superoxide than their normocholesterolemic counterparts.[4] Furthermore, SOD has been shown to be effective in attenuating the leukocyte-endothelial cell adhesion elicited by oxLDL in hamster dorsal skin fold arterioles and postcapillary venules,[17] and in rat mesenteric venules.[18] It has also been shown that placement of hamsters on an antioxidant diet containing vitamin C significantly reduces the leukocyte-endothelial cell adhesion and leukocyte-platelet aggregation elicited by oxLDL.[19]

The significance of an elevated production of superoxide by endothelial cells and/or leukocytes in hypercholesterolemic and atherosclerotic animals may lie in the ability of NO to chemically interact with superoxide. Since NO reacts with superoxide three times faster than SOD is able to dismutate superoxide,[20] it has been proposed that NO is a physiologically relevant scavenger of superoxide. The avidity of superoxide for NO (and vice versa) makes it difficult to discern whether the excess generation of superoxide observed in atherogenic vessels reflects an enhanced enzymatic production (e.g. via endothelial xanthine oxidase or neutrophilic NADPH oxidase) of the free radical or if it merely reflects a decline in endothelial cell NO production and a subsequent limitation of NO to scavenge basally synthesized superoxide. This interdependence between the rates of accumulation of superoxide and NO within the vasculature is also manifested at the enzymatic level. Recently, it was demonstrated that NO exerts a tonic and reversible inhibitory influence on xanthine oxidase activity through an alteration of the flavin prosthetic site.[21] Hence, one might predict that a reduction in NO synthesis in atherosclerotic vessels could result in an elevated accumulation of superoxide via both an attenuated capacity of cells to scavenge superoxide and due to release of the tonic inhibitory influence of NO on xanthine oxidase, which generates both superoxide and hydrogen peroxide. Support for this prediction is provided by recent reports[22,23] which demonstrate that inhibition of NOS with analogs of L-arginine results in a profound increase in oxidant generation by rat mesenteric venules.

Irrespective of the source and mechanism of oxygen radical production in HCh and atherosclerotic vessels, it is likely that the tendency to generate excessive quantities of oxygen radicals contributes to some of the microvascular alterations observed in these conditions. Oxygen radicals have been implicated as mediators of the leukocyte-endothelial cell adhesion, platelet-leukocyte aggregation, and vascular protein leakage associated with different acute and chronic inflammatory conditions.[22] Furthermore, superoxide and hydrogen peroxide appear to contribute to the modulation of endothelial cell adhesion molecules such as ICAM-1[24] and P-selectin,[25] as well as the formation and/or activation of some inflammatory mediators, such as platelet activating factor (PAF)[25] and complement.[26] Thus, one can readily envision how the abnormally low output of NO that accompanies atherosclerosis would favor an excessive generation and/or accumulation of ROM, which in turn elicit the formation of inflammatory mediators, enhance the expression of endothelial cell adhesion molecules, promote leukocyte-endothelial cell adhesion, and induce endothelial cell barrier dysfunction (increased albumin leakage).

LEUKOCYTE-ENDOTHELIAL CELL ADHESION IN HYPERCHOLESTEROLEMIA AND ATHEROSCLEROSIS

The participation of leukocytes in the pathogenesis of atherosclerosis was first suggested by Anitschkow in the early 1930s.[27] Leukocyte adherence to aortic endothelium and residence within atherosclerotic lesions from human autopsy specimens was demonstrated by electron microscopy in the early 1960s.[28] Since then, leukocytes have been identified within atherosclerotic lesions in different animal models of atherosclerosis as well as in human specimens.[29,30] Although much attention has been devoted to the adhesion of monocytes and macrophages because of their apparent contribution to foam cell formation in atherosclerotic plaques, there is growing evidence that invokes a role for other leukocyte subpopulations, including neutrophils and lymphocytes, in this disease

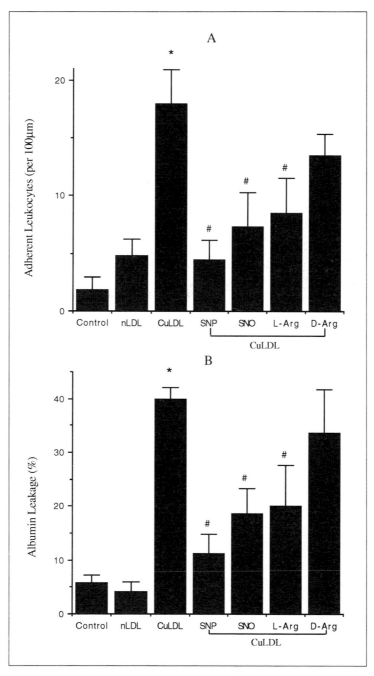

Fig. 6.1. Effects of NO donors on leukocyte adherence (A) and albumin leakage (B) in mesenteric venules exposed to CuLDL. *P < 0.05 relative to control, #P < 0.05 relative to CuLDL (n-6 in each group. Modified from ref. 18).

process. In experimental hypercholesterolemia-induced atherosclerosis, initial lesion formation is characterized by a prominent infiltration of monocytes and lymphocytes[29,31-34] with the number of adherent leukocytes correlated with the severity of lesion formation.[35] The presence of neutrophils in these plaques was identified only in the later stages of lesion development.[36-38]

There are several reports that invoke an altered leukocyte-endothelial cell adhesion response in animals with hypercholesterolemia. Diet-induced hypercholesterolemia in rats and rabbits is associated with an increased adhesion of leukocytes to aortic endothelium that begins as soon as one week and persists for up to one year of the diet.[29,39] A similar hyperadhesiveness of leukocytes has been noted in postcapillary venules of the ear in HCh rabbits.[40] Neutrophils isolated from HCh rabbits adhere more avidly to artificial surfaces in vitro than cells isolated from their normocholesterolemic counterparts, suggesting that the adhesion defect occurs in the leukocyte. A similar role for endothelial cells in this hyperadhesive response is supported by the observation that neutrophils isolated from normal rabbits bind more intensely to coronary artery rings from cholesterol-fed rabbits than to rings isolated from normal rabbits.[41]

In various animal models of atherogenesis, focal areas of predilection for development of plaques can be identified before such lesions are visible microscopically.[42-45] These predisposed areas have been shown to exhibit increased endothelial permeability with subsequent subendothelial accumulation of plasma proteins[46] and low density lipoproteins (LDL).[47,48] Not unexpectedly, leukocyte adhesion and subendothelial accumulation develops in these areas of altered endothelial integrity.[49] Thus, the adhesion of leukocytes to arterial endothelium in these prelesional areas is seen as one of the earliest events in atherosclerosis.[1] Based upon available information, it appears that a causal relationship exists between the initiation of leukocyte adhesion in atherogenesis and endothelial dysfunction induced by hypercholesterolemia.

The earliest identifiable event in atherogenesis is the focal accumulation of LDL (particularly apolipoprotein B-containing lipoproteins) within the intima, which occurs before endothelial permeability increases.[50-52] However, clusters of adherent leukocytes are observed on the surface of aortic endothelium within 1-2 weeks following dietary induction of high level hypercholesterolemia in

swine[49] and rabbits.[33] Adherent leukocytes migrate transendo-
thelially, monocytes differentiate into macrophages, and phagocy-
tose oxLDL thus becoming foam cells.[53] The accumulation and
lipid uptake by these cells results in fatty streak formation, the
first stage in atherosclerotic lesion development. Such areas expand
under continued hyperlipidemic conditions showing further monocyte recruitment and attachment at the lesional borders.[1,54] Thus,
based upon the available data in the literature it appears that the
induction of atherosclerosis is dependent upon two key processes:
1) enhanced transcytosis and accumulation of plasma proteins and
oxLDL within the subendothelial space of large blood vessels; and
2) chemotactic recruitment and adhesion of monocytes to the over-
lying endothelial surface with their subsequent emigration.[55]

ROLE OF PLASMA LIPIDS

The "lipid hypothesis" of atherosclerosis suggests that circulating cholesterol, primarily in the form of LDL, is fundamental to
atherogenesis such that atherosclerosis will not occur in the absence of significantly elevated levels.[56,57] Indeed, it has been shown
both in vivo and in vitro that monocyte adhesion to arterial endothelial cells is correlated to LDL levels.[35,58]

The evidence suggesting that oxidative modification of LDL
within the vascular wall[56,59] or bloodstream[60,61] is a fundamental
event in the induction of atherogenesis[56] has led to the "oxidative
modification hypothesis" of atherogenesis.[62] Lending support to this
hypothesis is the relationship between LDL susceptibility to oxidation and severity of atherosclerotic lesions in humans.[63] The presence of oxLDL within atherosclerotic plaques[56,64,65] and even the
bloodstream[60,61] is well documented, although, under most circum-
stances plasma lipids are likely protected from oxidation by circulating antioxidants. The effects of oxLDL are complex and have
been invoked in different aspects of atherogenesis, including leukocyte recruitment, endothelial cell dysfunction, subendothelial
trapping of macrophages, their transformation into foam cells, altered vasoregulation, and cytokine/growth factor generation.[56]

It has been suggested that LDL oxidation occurs when endothelial cell- and/or macrophage-derived oxygen free radicals over-
whelm the limited antioxidant capacity of affected vascular tissues.[66]
In early atherosclerotic lesions, the degree of LDL oxidation is likely
small due to the relative acellularity of the intima at this stage.[67,68]

Limited oxidation of LDL in these early lesions may result in the formation of minimally modified LDL (MM-LDL) which has been shown to have important atherogenic effects.[68] MM-LDL particles are still recognized by the specific low capacity (normal) LDL receptor[67] unlike the oxLDL particle which is only recognized by the high capacity scavenger (acetyl-LDL) receptor.[68] MM-LDL incubation stimulates monocyte but not PMN binding to cultured large vessel endothelial cells.[67] Additionally, MM-LDL induces endothelial cells to secrete colony stimulating factors[69] and monocyte chemotactic protein-1 (MCP-1)[68,70] while inhibiting production of platelet derived growth factor (PDGF).[71] Thus, it appears that MM-LDL may be active in initiating monocyte recruitment by endothelial cells of prelesional vessels. With the accumulation and activation of macrophages and resident vascular cells (endothelial cells and smooth muscle cells) more substantial oxidation of LDL, especially apolipoprotein B, occurs and results in a self-sustaining process provided HCh conditions persist.[67]

Oxidation of LDL in vitro has been shown to occur by the initiation of peroxidation reactions within the LDL particle.[72] These reactions are promoted by copper ions, heme proteins, and organic peroxyl radicals. Other studies have shown that in the presence of transition metals, LDL can also be oxidized by three major cell types of the atherogenic vessel wall: macrophages;[66,73] endothelial cells;[74,75] and smooth muscle cells.[74,76] The ability of neutrophils to cause LDL oxidation remains somewhat controversial although a recent report demonstrated that mild LDL oxidation occurs in vitro depending on the neutrophil activator utilized.[77] The mechanisms responsible for LDL oxidation in vivo are less well understood. However, it appears that the oxidation process requires the generation and breakdown of lipid hydroperoxide groups,[78] which (in the presence of transition metals) may lead to a chain reaction of unsaturated fatty acid peroxidation within the LDL particle.[79,80]

The oxidation of LDL appears to be mediated by the lipoxygenase pathway[81-84] and/or by superoxide anion-dependent[85-88] processes. It has been hypothesized that lipoxygenase inserts hydroperoxide groups into unsaturated fatty acid side chains within LDL particles.[80] The degradation of the hydroperoxide groups appears to be catalyzed by transition metal ions either free or bound to heme groups.[78] The existence of superoxide-mediated LDL oxidation is based upon in vitro studies showing that administration of SOD inhibits LDL oxidation[77,86] and that the degree of LDL modification is correlated to superoxide anion production.[77] However, another investigation has shown no alteration in the level of LDL oxidation by macrophages when superoxide anion production was increased 5-10 times by NADPH oxidase stimulation.[79] Thus, it has been suggested that the role of the superoxide anion in LDL modification is questionable and that the protective effects of SOD are actually due to the metal ion chelating properties of SOD in transition metal-dependent oxidation studies.[79]

Experimental studies have demonstrated that administration of oxLDL causes leukocyte adhesion both in vivo and in vitro.[89-93] The chemotactic activity of oxLDL both alone[94] and in the presence of cultured endothelial cells[67] is well documented for monocytes. Although cultured endothelial cells constitutively produce the chemotaxin, MCP-1,[95] exposure of oxLDL in vitro stimulates further release of MCP-1 from endothelial cells and macrophages[95-98] and has been identified in macrophage derived foam cells within atherosclerotic lesions.[98] This suggests that following their initial recruitment, monocytes may amplify their subsequent recruitment through MCP-1 release.[96] Similarly, in vitro MM-LDL administration induces endothelial cells to produce MCP-1[70] and colony stimulating factors (CSF)[69] while in vivo administration to mice resulted in elevated levels of JE, the mouse homologue for MCP-1, and macrophage-derived CSF.[68] Furthermore, monocytes exposed to oxLDL in vitro produce interleukin-8 (IL-8), a T lymphocyte chemotaxin with the IL-8 production being proportional to the degree of LDL oxidation.[99] IL-8 which can be produced by numerous cell types in response to interleukin-1 (IL-1) stimulation[100] is also chemotactic for neutrophils but requires levels two to ten times higher than the monocyte.[101]

Inherent chemotactic activity also exists within the lipid fraction of the oxLDL particle and has been proposed to be due to lysophosphatidylcholine (lysoPC).[94,102] Increased levels of lysoPC are found within diet-induced atherosclerotic lesions[103] as well as in oxLDL.[102,104,105] LysoPC has been shown to be responsible for the chemotactic response to oxLDL by both monocytes and T-lymphocytes.[102,106] Once monocytes have been chemotactically recruited and have differentiated into resident macrophages, oxLDL inhibits chemotaxis,[107] thus serving as a mechanism to maintain the leukocytic infiltrate within the lesion.

In recent years, it has been shown that exogenously administered oxLDL promotes the adhesion of leukocytes within microvessels of the mouse,[108] rat,[18] and hamster.[90] In mouse and hamsters, systemic administration of oxLDL elicits leukocyte rolling and firm adhesion in arterioles and postcapillary venules of skeletal muscle.[90-92,108] In the rat mesentery, both oxLDL and MM-LDL[93] as well as oxidized chylomicrons[109] promote leukocyte-endothelial cell adhesive interactions that are limited to venules. In an elegant series of studies, Lehr and associates have obtained evidence that implicates ROM, leukotrienes and PAF in the leukocyte-endothelial cell adhesion elicited by oxLDL.[17,90,92]

Evidence for superoxide anion-dependent, oxLDL-mediated leukocyte adhesion in the microcirculation is based upon studies that demonstrate that SOD attenuates leukocyte adhesion induced by oxLDL[17] and cigarette smoke (CS)[110] in hamsters. The source of superoxide anions in the CS studies is unknown. CS has been shown to increase xanthine oxidase activity in vivo, however, the leukocyte adhesion elicited by CS is not blocked by allopurinol, an inhibitor of xanthine oxidase.[110] CS has been shown to markedly enhance leukocyte production of superoxide anions in vitro and therefore phagocytes may be a source of the superoxide.[111] As mentioned previously, the elevated levels of superoxide may also result secondarily from an oxLDL-induced attenuation of NO production.

Oxidized LDL acts as a calcium ionophore to elevate intracellular calcium levels and consequently results in an activation of the lipoxygenase pathway.[112] In vivo studies in mice and hamsters have demonstrated that oxLDL-induced leukocyte adhesion is completely blocked by administration of a selective lipoxygenase inhibitor, MK-886.[90,113] Leukocyte adhesion but not leukocyte rolling, was inhibited by MK-886 in a CS-induced hamster model implying that adhesion and rolling are likely mediated by different adhesion molecules.[113] Further support for a role of leukotrienes in atherogenesis is the detection of increased leukotriene biosynthesis within atherosclerotic plaques.[114] In addition to activating leukocyte adhesion receptors, leukotrienes, especially leukotriene B_4 (LTB$_4$), are potent chemoattractants, thereby allowing these lipid mediators to enhance leukocyte infiltration through multiple mechanisms.

PAF and PAF-like lipids (PAF-LL) which are oxidatively fragmented phospholipids with short *sn2* polyunsaturated fatty acid residues activate a specific PAF receptor on leukocytes.[115] Synthesis of PAF occurs by enzymatic production from membrane phospholipids while PAF-LL are created by unrestrained ROM peroxidation of membrane phospholipids.[115,116] It has been theorized that oxLDL works through membrane lipid peroxidation by ROM to produce PAF and PAF-LL.[92] Occupation of the PAF receptor on monocytes and neutrophils results in both a potent chemotactic response and leukocyte adhesion to endothelial cells.[117] Further support for the role of PAF in atherosclerotic leukocyte adhesion is provided by the demonstration that a PAF receptor antagonist, WEB2170, attenuates oxLDL-induced leukocyte adhesion, but not rolling, in postcapillary venules.[92]

The rapidity of the leukocyte-endothelial cell adhesion responses observed in postcapillary venules exposed to oxLDL are consistent with a role for PAF and LTB_4. Both of these lipid mediators of inflammation are known to elicit the rapid surface expression of β_2-integrins on leukocytes,[118] which can promote leukocyte adhesion with constitutively expressed ICAM-1 on the surface of endothelial cells. Indeed, using monoclonal antibodies (MAbs) directed against different adhesion epitopes on leukocytes (CD11/CD18), Lehr and colleagues[108] have demonstrated that the recruitment of firmly adherent leukocytes in mice dorsal skin arterioles and postcapillary venules exposed to oxLDL is largely dependent on the upregulation of CD11/CD18 on leukocytes. Recent studies in our lab showed that the oxLDL-induced leukocyte adherence and emigration, and enhanced albumin leakage in rat mesenteric postcapillary venules, were significantly attenuated by pretreatment with MAbs directed against CD11/CD18, P-selectin, and L-selectin, but not E-selectin. These results suggest that the oxLDL-induced leukocyte adhesion and emigration are mediated by CD11/CD18 and L-selectin on activated neutrophils and P-selectin on venular endothelium.

While adhesion molecule-specific MAbs have proven to be very helpful in identifying the molecular determinants of oxLDL-induced leukocyte adhesion, other reagents have provided some insight into the factors that may contribute to the modulation of leukocyte and/or endothelial cell adhesion molecule expression elicited by oxLDL. One group of reagents that has proven to be very effective is compounds that act to generate NO at the level of the microvasculature. A role for abnormal NO metabolism (decreased production and/or increased inactivation by superoxide) as a causative factor in microvascular dysfunction elicited by oxLDL is supported by several lines of evidence: 1) endothelial cell-derived NO is inactivated by oxLDL or superoxide,[11,119] 2) the constitutive and inducible forms of NOS are both inactivated by ox-LDL,[13] 3) inhibitors of NO production elicit leukocyte-endothelial cell adhesion in otherwise normal postcapillary venules,[120] and agents that donate NO have been shown to attenuate the leukocyte adhesion observed under conditions of enhanced superoxide formation, such as ischemia/reperfusion.[120] Hence, one might expect that if oxLDL produces an imbalance between the NO and superoxide produced by endothelial cells, then the resultant leukocyte adhesion response and any accompanying microvascular dysfunction may be ablated either by the administration of NO donors or by supplementation with L-arginine, the substrate for endothelial cell production of NO.

NO donors have been employed in several studies in an attempt to determine whether an elevation or restoration of reduced levels of NO alter the microvascular and/or tissue responses to different experimental perturbations. NO donors appear to be effective in attenuating the leukocyte-endothelial cell adhesion elicited by either platelet activating factor,[121] ischemia/reperfusion,[120,122] or NOS inhibitors.[123] Similarly, NO donors have also been shown to reduce the mast cell degranulation associated with ischemia/reperfusion[120] and inhibition of NOS,[124] as well as the albumin leakage elicited by ischemia/reperfusion.[120]

L-arginine supplementation is another strategy that has been used to assess the potential contribution of NO to different pathological processes.[5,39] Since NO is derived from the metabolism of L-arginine, providing an exogenous source of this amino acid has been employed as a means of enhancing, or restoring to normal, the production of NO. In some model systems, L-arginine supplementation offers no benefit,[125] while it is highly protective in others. For example, it has been shown that acute administration of L-arginine to HCh animals or humans normalizes NO dependent vasodilation[5] as well as the enhanced endothelial cell adhesiveness for monocytes.[39]

Lefer and Ma[41] have demonstrated that hypercholesterolemia reduces NO release and causes a corresponding increase in neutrophil adherence to rabbit coronary artery endothelium. The increased leukocyte-endothelial cell adhesion appeared to result from enhanced endothelial cell adhesiveness rather than to changes in the properties of the leukocytes. Both the decreased production of NO and the increased neutrophil adhesion could be reversed by addition of L-arginine to isolated coronary arteries. Similar attenuations of the altered NO production and enhanced neutrophil adhesion responses were observed in rabbits receiving the 3-hydroxy-3-methylglutaryl coenzyme A reductase inhibitor, lovastatin. Since lovastatin treatment also resulted in a profound reduction in plasma cholesterol concentration, the latter findings tend to implicate some form of circulating cholesterol as a factor eliciting the decline in NO production by vascular endothelial cells.

In a recent study,[18] we have examined whether NO donors and L-arginine attenuate oxLDL-induced increases in leukocyte adhesion, venular albumin leakage, and mast cell degranulation in rat mesentery. Leukocyte rolling, adherence, and emigration and the leakage of FITC-albumin were monitored in postcapillary venules while mast cell degranulation was determined in the adjacent extravascular compartment of rat mesentery prior to and during local intra-arterial infusion of either normal LDL (n-LDL) or copper-oxidized LDL (Cu-LDL). Infusion of Cu-LDL, but not n-LDL, caused significant increases in leukocyte rolling, adherence, and emigration, mast cell degranulation, and an enhanced albumin leakage response. Simultaneous superfusion of the mesenteric microcirculation with either the NO donors sodium nitroprusside, spermine-NO, or with L-arginine significantly reduced the Cu-LDL-induced leukocyte adherence/emigration, mast cell degranulation, and albumin leakage, whereas D-arginine had no effect. The recruitment of rolling leukocytes elicited by Cu-LDL was significantly reduced only by spermine-NO. These results indicate that all of the deleterious microvascular alterations elicited by Cu-LDL are profoundly reduced by agents that either release or enhance the production of nitric oxide.

ROLE OF CIRCULATING INHIBITORS OF NO SYNTHASE

Most of the available information concerning the physiologic and pathologic roles of NO is based on experimentation involving the use of synthetic analogs of L-arginine (e.g., N^G-nitro-L-arginine methyl ester, L-NAME) that act to inhibit NO synthesis from

L-arginine. Recent studies indicate that methylated L-arginine derivatives such as N^G, $N^{G'}$-dimethyl-L-arginine (L-DMA) are normally present in the plasma of humans and rats.[126] Guanidino-N-methylated arginines such as L-DMA are widely distributed in proteins that are methylated post-translationally, and these methylated arginines are continuously released into extracellular fluid as a consequence of protein breakdown.[127] Substantial amounts of free L-DMA normally exist in certain tissues. For example, it has been estimated that the kidney contains 35 μM L-DMA,[127] suggesting that this organ and other tissues may represent a large pool for release of the NO synthase inhibitor into extracellular fluid.

While the plasma concentration of L-DMA in normal human subjects and rabbits is approximately 1 μM, plasma levels rise to about 10 μM in rabbits with hypercholesterolemia.[10] This elevated level of L-DMA measured in HCh rabbits appears sufficient to significantly inhibit NO production by murine macrophages (J774 cell-line) and rat aortic endothelial cells.[128] These observations suggest that an accumulation of L-DMA in the extravascular space may ex-

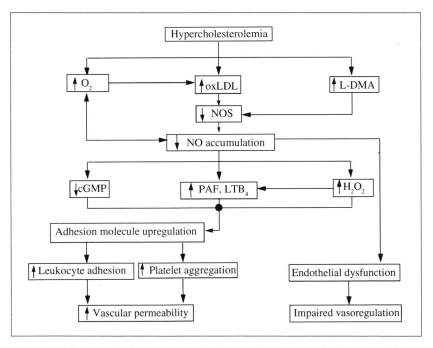

Fig. 6.2. Schematic representation of the mechanisms involved in modulating leukocyte adhesion in atherosclerosis. O_2^-: superoxide anion; cGMP: cyclic guanosine monophosphate; H_2O_2: hydrogen peroxide.

QUESTIONNAIRE

Receive a FREE BOOK of your choice

Please help us out—Just answer the questions below, then select the book of your choice from the list on the back and return this card.

R.G. Landes Company publishes five book series: *Medical Intelligence Unit, Molecular Biology Intelligence Unit, Neuroscience Intelligence Unit, Tissue Engineering Intelligence Unit* and *Biotechnology Intelligence Unit.* We also publish comprehensive, shorter than book-length reports on well-circumscribed topics in molecular biology and medicine. The authors of our books and reports are acknowledged leaders in their fields and the topics are unique. Almost without exception, there are no other comprehensive publications on these topics.

Our goal is to publish material in important and rapidly changing areas of bioscience for sophisticated scientists. To achieve this goal, we have accelerated our publishing program to conform to the fast pace in which information grows in bioscience. Most of our books and reports are published within 90 to 120 days of receipt of the manuscript.

Please circle your response to the questions below.

1. We would like to sell our *books* to scientists and students at a deep discount. But we can only do this as part of a prepaid subscription program. The retail price range for our books is $59-$99. Would you pay $196 to select four *books* per year from any of our Intelligence Units–$49 per book–as part of a prepaid program?

           Yes   **No**

2. We would like to sell our *reports* to scientists and students at a deep discount. But we can only do this as part of a prepaid subscription program. The retail price range for our reports is $39-$59. Would you pay $145 to select five *reports* per year–$29 per report–as part of a prepaid program?

           Yes   **No**

3. Would you pay $39–the retail price range of our books is $59-$99–to receive any single book in our Intelligence Units if it is spiral bound, but in every other way identical to the more expensive hardcover version?

           Yes   **No**

To receive your free book, please fill out the shipping information below, select your free book choice from the list on the back of this survey and mail this card to:

R.G. Landes Company, 909 S. Pine Street, Georgetown, Texas 78626 U.S.A.

Your Name _____

Address _____

City_____ State/Province:_____

Country:_____ Postal Code:_____

My computer type is Macintosh_____ ; IBM-compatible _____ ; Other _____

Do you own ____ or plan to purchase ___ a CD-ROM drive?

AVAILABLE FREE TITLES

Please check three titles in order of preference.
Your request will be filled based on availability. Thank you.

☐ Water Channels
Alan Verkman,
University of California-San Francisco

☐ The Na,K-ATPase:
Structure-Function Relationship
J.-D. Horisberger, University of Lausanne

☐ Intrathymic Development of T Cells
J. Nikolic-Zugic,
Memorial Sloan-Kettering Cancer Center

☐ Cyclic GMP
Thomas Lincoln, University of Alabama

☐ Primordial VRM System and the Evolution
of Vertebrate Immunity
John Stewart, Institut Pasteur-Paris

☐ Thyroid Hormone Regulation
of Gene Expression
Graham R. Williams, University of Birmingham

☐ Mechanisms of Immunological Self Tolerance
Guido Kroemer, CNRS Génétique Moléculaire et
Biologie du Développement-Villejuif

☐ The Costimulatory Pathway
for T Cell Responses
Yang Liu, New York University

☐ Molecular Genetics of Drosophila Oogenesis
Paul F. Lasko, McGill University

☐ Mechanism of Steroid Hormone Regulation
of Gene Transcription
M.-J. Tsai & Bert W. O'Malley, Baylor University

☐ Liver Gene Expression
François Tronche & Moshe Yaniv,
Institut Pasteur-Paris

☐ RNA Polymerase III Transcription
R.J. White, University of Cambridge

☐ src Family of Tyrosine Kinases in Leukocytes
Tomas Mustelin, La Jolla Institute

☐ MHC Antigens and NK Cells
Rafael Solana & Jose Peña,
University of Córdoba

☐ Kinetic Modeling of Gene Expression
James L. Hargrove, University of Georgia

☐ PCR and the Analysis of the T Cell Receptor
Repertoire
Jorge Oksenberg, Michael Panzara & Lawrence
Steinman, Stanford University

☐ Myointimal Hyperplasia
Philip Dobrin, Loyola University

☐ Transgenic Mice as an In Vivo Model
of Self-Reactivity
David Ferrick & Lisa DiMolfetto-Landon,
University of California-Davis and Pamela Ohashi,
Ontario Cancer Institute

☐ Cytogenetics of Bone and Soft Tissue Tumors
Avery A. Sandberg, Genetrix & Julia A. Bridge ,
University of Nebraska

☐ The Th1-Th2 Paradigm and Transplantation
Robin Lowry, Emory University

☐ Phagocyte Production and Function Following
Thermal Injury
Verlyn Peterson & Daniel R. Ambruso,
University of Colorado

☐ Human T Lymphocyte Activation Deficiencies
José Regueiro, Carlos Rodríguez-Gallego
and Antonio Arnaiz-Villena,
Hospital 12 de Octubre-Madrid

☐ Monoclonal Antibody in Detection and
Treatment of Colon Cancer
Edward W. Martin, Jr., Ohio State University

☐ Enteric Physiology of the Transplanted Intestine
Michael Sarr & Nadey S. Hakim, Mayo Clinic

☐ Artificial Chordae in Mitral Valve Surgery
Claudio Zussa, S. Maria dei Battuti Hospital-Treviso

☐ Injury and Tumor Implantation
Satya Murthy & Edward Scanlon,
Northwestern University

☐ Support of the Acutely Failing Liver
A.A. Demetriou, Cedars-Sinai

☐ Reactive Metabolites of Oxygen and Nitrogen
in Biology and Medicine
Matthew Grisham, Louisiana State-Shreveport

☐ Biology of Lung Cancer
Adi Gazdar & Paul Carbone,
Southwestern Medical Center

☐ Quantitative Measurement
of Venous Incompetence
Paul S. van Bemmelen, Southern Illinois University
and John J. Bergan, Scripps Memorial Hospital

☐ Adhesion Molecules in Organ Transplants
Gustav Steinhoff, University of Kiel

☐ Purging in Bone Marrow Transplantation
Subhash C. Gulati,
Memorial Sloan-Kettering Cancer Center

☐ Trauma 2000: Strategies for the New Millennium
David J. Dries & Richard L. Gamelli,
Loyola University

plain, at least in part, the microvascular dysfunction and cell-cell adhesive interactions (leukocyte-endothelial cell; leukocyte-platelet) that is observed in pathological conditions that are characterized by impaired production of NO by vascular endothelial cells (e.g., hypercholesterolemia). The potential significance of L-DMA as an endogenous modulator of NO production by the microvasculature is further supported by studies that demonstrate the release of L-DMA by human umbilical vein endothelial cells in amounts that are sufficient to inhibit NO production by cocultured macrophages (J774 cell-line).[129] These observations suggest that L-DMA may exert an autocrine influence on NO production by endothelial cells and by parenchymal cells in certain tissues (e.g., kidney).

In view of the mounting evidence that L-DMA may act as an endogenous modulator of NO production, we undertook a series of experiments to determine whether L-DMA, at concentrations measured in extracellular fluid of rabbits with hypercholesterolemia,[10] affects the microvasculature in a manner similar to that previously reported for synthetic analogs of L-arginine that inhibit NO production.[130] Leukocyte adherence and emigration, leukocyte-platelet aggregation, and the extravasation of albumin were monitored in rat mesenteric venules exposed to different concentrations (1-100 μM) of either L-DMA or L-NAME. Increases in leukocyte adherence (7- to 9-fold) and emigration (3- to 5-fold), platelet-leukocyte aggregation, mast cell degranulation and an enhanced albumin leakage (30-50%) were observed within 30 mins after exposing the microvascular bed to either inhibitor, however, leukocyte emigration and albumin leakage responded more intensely to L-NAME than to L-DMA. The microvascular alterations and mast

Table 6.1. Dose-dependent microvascular alteration elicited by either L-NAME or L-DMA

	Control	L-NAME 10 μM	L-NAME 100 μM	L-DMA 10 μM	L-DMA 100 μM
Leukocyte adherence	2.5 ± 0.7	7.6 ± 0.7[a]	17.0 ± 1.5[a]	9.0 ± 1.8[a]	13.4 ± 1.6[a]
Albumin leakage	4.7 ± 1.9	21.9 ± 3.4[a]	55.6 ± 6.3[a]	16.7 ± 5.6	29.7 ± 5.5[a]
Platelet-leukocyte aggregates	0	2.8 ± 1.0[a]	6.1 ± 0.4[a]	2.6 ± 1.4	5.8 ± 2.3[a]
Degranulated mast cells	2.8 ± 1.3	19.3 ± 5.4[a]	33.9 ± 4.0[a]	27.3 ± 9.0[a]	19.6 ± 5.0[a]

L-NAME: N^G-nitro-L-arginine ester, L-DMA: $N^G,N^{G'}$-dimethyl-L-arginine.
[a]Significant difference (P < 0.05) from corresponding control value.
From Ref. 130.

cell degranulation elicited by the two inhibitors were attenuated by addition of L-arginine to the superfusate. These results suggest that the endogenous NOS inhibitor (L-DMA) is capable of eliciting an inflammatory response at concentrations detected in plasma of rabbits with hypercholesterolemia.

Although L-DMA is an endogenous inhibitor of NO production that is normally synthesized by vascular endothelial cells[129] and accumulates in extracellular fluid during certain disease states,[10,128] relatively little is known about its biological actions. Gardiner and associates[131] have recently reported that L-DMA and L-NMMA are equipotent inhibitors of NOS based on similarities of the renal, mesenteric, and hindquarter vasoconstrictor responses to the two L-arginine derivatives. Vallance et al[128] reached a similar conclusion when they compared the inhibitory effects of L-DMA and L-NMMA on NO synthesis by J774 cells, which exhibit the inducible form of NOS. We have compared the abilities of L-DMA and L-NAME to inhibit NOS activity in rat mesentery (primarily the constitutive form) and in rat neutrophils (primarily the inducible form).[130] While L-DMA did exhibit an inhibitory effect on NOS in both tissues, L-DMA proved to be a much less potent inhibitor of the enzyme in our in vitro assay system. Our observation that 10 μM L-DMA did not significantly inhibit NOS in homogenates of mesenteric tissue or neutrophils, while eliciting a number of microvascular responses that are L-arginine reversible, suggests that either intact cell transduction mechanisms are needed for L-DMA to exert its cellular actions or that L-DMA is rapidly metabolized by tissues to form a more potent NOS inhibitor. In rats, the enzyme dimethylargininase hydrolyzes L-DMA to form L-citrulline and dimethylamine (DMN), with most of the latter excreted by the kidneys.[127]

PLATELET-LEUKOCYTE AGGREGATION IN HYPERCHOLESTEROLEMIA AND ATHEROSCLEROSIS

Intravascular activation of platelets, with its attendant release of chemical mediators, may have important consequences in the pathogenesis of atherosclerosis. Platelets are known to interact with both circulating neutrophils and vascular endothelium during different inflammatory conditions. Many compounds released by activated neutrophils may act as platelet agonists (e.g., superoxide, hydrogen peroxide) or antagonists (ADPase); conversely, platelets can also release factors that may either inhibit (soluble P-selectin, NO) or activate (thromboxane A$_2$) neutrophils.[132] Platelets can also facilitate the formation of inflammatory mediators by endothelial cells through transcellular exchange of precursor metabolites. The formation of platelet-leukocyte aggregates has been observed on the endothelial cell surface of postcapillary venules exposed to NOS inhibitors,[133] cigarette smoke,[134] or oxLDL.[135] The rate of formation of these aggregates can be significantly attenuated by monoclonal antibodies directed against P-selectin, findings consistent with an interaction between P-selectin expressed on the surface of activated platelets and a counter-receptor (e.g., sialyl Lewis X) on leukocytes.[136] The aforementioned observations are consistent with the view that NO serves to inhibit the adhesion of platelets to endothelial cells as well as leukocytes.[120] Conditions that are associated with a reduction of NO production and/or an increased inactivation of NO by superoxide (e.g., hypercholesterolemia) would therefore be expected to elicit the mobilization of P-selectin to the surface of platelets, where it promotes platelet-platelet and leukocyte-platelet adhesive interactions.

The potential clinical significance of the platelet hyperreactivity and tendency for aggregation of platelets with leukocytes in hypercholesterolemia is exemplified by reports that there is a more profound accumulation of platelets and much larger infarct sizes after ischemia/reperfusion in hearts of HCh rabbits relative to their normocholesterolemic counterparts, and that rendering the hypercholesterolemic animals thrombocytopenic with antiplatelet serum significantly reduces infarct size, while having no beneficial effect in normal animals.[137] It remains unknown whether interventions that interfere with platelet-leukocyte aggregation (P-selectin antibodies, NO donors) or platelet-endothelial cell adhesion (e.g., antibodies to GPIIb/IIIa) also afford more protection against the ischemia/reperfusion injury in HCh animals than in their normocholesterolemic counterparts.

CONCLUSIONS

The growing body of evidence that implicates immune cells in the pathogenesis of atherosclerosis readily justifies the need for a better understanding of the mechanisms that initiate and sustain the recruitment of these cells into the vascular wall. There is ample support in the literature for the concept that a decreased production and/or increased inactivation of NO contributes to the recruitment and activation of inflammatory cells observed in hypercholesterolemia and atherosclerosis. Additional work is needed to assess the relative contributions of oxidized lipoproteins and endogenous inhibitors of NOS to the pathogenesis of this disease. Rapid advances in this field are likely to result from the recent development of genetically-engineered animal models that either lack or overexpress these putative mediators. While the biochemical and molecular basis for the linkage between abnormal NO metabolism and elevated blood cholesterol levels remain undefined, the beneficial actions of NO donors and L-arginine in animal models of atherosclerosis lend credence to the strategy of developing new therapeutic agents that are targeted at minimizing the decline in NO levels in atherosclerotic blood vessels.

REFERENCES

1. Ross R. The pathogenesis of atherosclerosis-an update. N Engl J Med 1986; 314;488-500.
2. Garg UC, and Hassid A. Nitric oxide-generating vasodilators and 8-bromo-cyclic guanosine monophosphate inhibit mitogenesis and proliferation of cultured rat vascular smooth muscle cells. J Clin Invest 1989; 83:1774-7.
3. Guerra RJ, Brotherton AFA, Goodwin PJ et al. Mechanisms of abnormal endothelium-dependent vascular relaxation in atherosclerosis: Implications for altered autocrine and paracrine function of EDRF. Blood Vessels 1989; 26:300-14.
4. Harrison DG. Endothelial dysfunction in atherosclerosis. Basic Research in Cardiology 89: 87-102, 1994.
5. Creager MA, Gallagher SJ, Girerd XJ et al. L-arginine improves endothelium-dependent vasodilation in hypercholesterolemic humans. J Clin Invest 1992; 90:1248-53.
6. Schuschke DA, Joshua IG, Miller FN. Comparison of early microcirculatory and aortic changes in hypercholesterolemic rats. Arterioscler Thromb 1991; 11:154-60.
7. Naruse K, Shimizu K, Muramatsu M et al. Long-term inhibition of NO synthesis promotes atherosclerosis in the hypercholesterolemic rabbit thoracic aorta. PGH2 does not contribute to impaired endothelium-dependent relaxation. Arterioscler Thromb 1994; 14:746-52.
8. Walter R, Schaffner A, Blau N et al. Tetrahydrobiopterin is a secretory product of murine vascular endothelial cells. Biochem Biophys Res Comm 1994; 203:1522-6.
9. Randall MD, Ujie H, Griffith TM. L-arginine reverses the impairment of nitric oxide-dependent collateral perfusion in dietary-induced hypercholesterolemia in the rabbit. Clin Sci 1994; 87:53-9.

10. Yu X, Li Y, Xiong Y. Increase of an endogenous inhibitor of nitric oxide synthesis in serum of high cholesterol fed rabbits. Life Sciences 1994; 54:753-8.

11. Chin JH, Azhar S, Hoffman BB. Inactivation of endothelial derived relaxing factor by oxidized lipoproteins. J Clin Invest 1992; 89:10-8.

12. Kanwar S, Tepperman BL, Payne D et al. Time course of nitric oxide production and epithelial dysfunction during ischemia/reperfusion of the feline small intestine. Circ Shock 1994; 42:135-40.

13. Yang X, Cai B, Sciacca RR et al. Inhibition of inducible nitric oxide synthase in macrophages by oxidized low-density lipoproteins. Circ Res 1994; 74:318-28.

14. Raij L, Nagy K, Coffee K et al. Hypercholesterolemia promotes endothelial dysfunction in vitamin E-and selenium-deficient rats. Hypertension 1993; 22:56-61.

15. Mantha SV, Prasad M, Kalra J et al. Antioxidant enzymes in hypercholesterolemia and effects of vitamin E in rabbits. Atherosclerosis 1993; 101:135-44.

16. Mugge A, Edwell JH, Peterson TE et al. Chronic treatment with polyethylene-glycolated superoxide dismutase partially restores endothelium dependent vascular relaxations in cholesterol-fed rabbits. Circ Res 1991; 69:1293-1300.

17. Lehr HA, Becker M, Marklund SL et al. Superoxide-dependent stimulation of leukocyte adhesion by oxidatively modified LDL in vivo. Arterioscler Thromb 1992; 12:824-9.

18. Liao L, and Granger DN. Modulation of oxidized low density lipoprotein-induced microvascular dysfunction by nitric oxide. Am J Physiol 1995; 268:H1643-50.

19. Lehr HA, Frei B, Olofsson M et al. Protection from oxidized LDL-induced leukocyte adhesion to micro- and macrovascular endothelium in vivo by vitamin C, but not by vitamin E. Circulation 1995; 91:1525-32.

20. Huie RE, Padnaja S. The reaction of NO with superoxide. Free Rad Res Comm 1993; 18:195-200.

21. Fukahori M, Ichimori K, Ishida H et al. Nitric oxide reversibly suppresses xanthine oxidase activity. Free Rad Res Comm 1994; 21:203-12.

22. Kurose I, Wolf R, Grisham MB et al. Microvascular responses to inhibition of nitric oxide production. Role of active oxidants. Circ Res 1995; 76:30-9.

23. Suematsu M, Tamatani T, Delano FA et al. Microvascular oxidative stress preceeding leukocyte activation elicited by in vivo nitric oxide suppression. Am J Physiol 1994; 266:H2410-5.

24. Lo SK, Janakidevi K, Lai L et al. Hydrogen peroxide-induced increase in endothelial adhesiveness is dependent on ICAM-1 activation. Am J Physiol 1993; 264:L406-12.

25. Gaboury JP, Anderson DC, and Kubes P. Molecular mechanisms involved in superoxide-induced leukocyte-endothelial cell interactions in vivo. Am J Physiol 1994; 266:H637-42.
26. Shingu M, Nonaka S, Nishimukai H et al. Activation of complement in normal serum by hydrogen peroxide and hydrogen peroxide-related oxygen radicals produced by activated neutrophils. Clin Exp Immunol 1992; 90:72-8.
27. Anitschkow N. *Experimental arteriosclerosis in animals.* New York: The MacMillan Company, 1933.
28. Still WJS, Marriott PR. Comparative morphology of the early atherosclerotic lesion in man and cholesterol-atherosclerosis in the rabbit. An electron microscopic study. J Atheroscler Res 1964; 4:373-86.
29. Joris J, Zand T, Nunnari JJ et al. Studies on the pathogenesis of atherosclerosis. I. Adhesion and emigration of mononuclear cells in the aorta of hypercholesterolemic rats. Am J Pathol 1983; 113:341-358.
30. Takebayashi S, Kubota I, Kamino A et al. Ultrastructural aspects of human atherosclerosis, the role of foam cells and modified smooth muscle cells. J Electron Microsc 1972; 21:301-13.
31. Faggiotto A, Ross R, Harker E. Studies of hypercholesterolemia in the nonhuman primate. I. Changes that lead to fatty streak formation. Arterioscler Thromb 1984; 4:323-40.
32. Gerrity RG. The role of the monocyte in atherogenesis. I. Transition of blood-borne monocytes into foam cells in fatty lesion. Am J Pathol 1981; 103:181-90.
33. Hansson GK, Seifert PS, Olsson G et al. Immunohistochemical detection of macrophages and T lymphocytes in atherosclerotic lesions of cholesterol-fed rabbits. Arterioscler Thromb 1991; 11:745-50.
34. Schwartz CJ, Sprague EA, Kelley JL et al. Aortic intimal monocyte recruitment in the normo and hypercholesterolemic baboon (Papio cynocephalus). An ultrastructural study: Implications in atherogenesis. Virchows Arch [Pathol Anat] 1985; 405:175-91.
35. Jerome WG, Lewis JC. Early atherogenesis in white carneau pigeons: I. Leukocyte margination and endothelial alterations at the celiac bifurcation. Am J Pathol 1984; 116:56-68.
36. Faggiotto A, Ross R. Studies of hypercholesterolemia in the nonhuman primate. II. Fatty streak conversion to fibrous plaque. Arterioscler Thromb 1984; 4:341-56.
37. Stary HC, Manilow MR. Ultrastructure of experimental coronary artery atherosclerosis in *Cynomolgus Macaques.* Atherosclerosis 1982; 43:151-75.
38. Trillo AA. The cell population of aortic fatty streaks in African green monkey with special reference to granulocytic cells. An ultrastructural study. Atherosclerosis 1982; 43:259-75.
39. Tsao PS, McEvoy LM, Drexler H et al. Enhanced endothelial adhesiveness in hypercholesterolemia is attenuated by L-arginine. Circulation 1994; 89:2176-82.

40. Asano M, Ohkubo A, Hirokawa A et al. Intravital-microscopic observations on the intravascular behavior of blood cell components during dietary-induced hyperlipidemia in the male rabbit. Biorheology 1988; 25:329-38.

41. Lefer AM, Ma XL. Decreased basal nitric oxide release in hypercholesterolemia increases neutrophil adherence to rabbit coronary artery endothelium. Arterioscler and Thromb 1993; 13:771-6.

42. Bell FP, Day AJ, Gent M et al. Differing patterns of cholesterol accumulation and ^{3}H-cholesterol influx in areas of the cholesterol fed pig aorta identified by Evans Blue dye. Exp Mol Pathol 1975; 22:366-75.

43. Day AJ, Bell FP, Schwartz CJ. Lipid metabolism in focal areas of normal-fed and cholesterol-fed pig aorta. Exp Mol Pathol 1974; 21:179-93.

44. Fry DL. Responses of the arterial wall to certain physical factors. In: Atherogenesis: Initiating Factors. Ciba Foundation Symposium 12. Elsevier, Amsterdam, London, New York 93, 1973.

45. McGill HC, Geer JC, Holman RL. Sites of vascular vulnerability in dogs demonstrated by Evans Blue. AMA Arch Pathol 1957; 64:303-11.

46. Bell FP, Adamson I, Schwartz CJ. Aortic endothelial permeability to abumin: Focal and regional patterns of uptake and transmural distribution of ^{125}I-albumin in the young pig. Exp Mol Pathol 1974; 20:57-68.

47. Feldman DL, Hoff HF, Gerrity RG. Immunohistochemical localization of Apo B in aortas from hyperlipidemic swine. Preferential accumulation in lesion-prone areas. Arch Pathol Lab Med 1984; 108:817-22.

48. Hoff HF, Gerrity RG, Naito HK et al. Quantitation of Apo B in aortas of hypercholesterolemic swine. Lab Invest 1983; 48:492-504.

49. Gerrity RG, Naito HK, Richardson M et al. Dietary-induced atherogenesis in swine. I. Morphology of the intima in prelesion stages. Am J Pathol 1979; 95:775-92.

50. Frank JS, Fogelman AM. Ultrastructure of the intima in WHHL and cholesterol-fed rabbit aortas prepared by ultra-rapid freezing and freeze-etching. J Lipid Res 1989; 30:967-78.

51. Schwenke DC, Carew TE. Initiation of atherosclerotic lesions in cholesterol-fed rabbits. I. Focal increases in arterial LDL concentration precede development of fatty streak lesions. Arteriosclerosis 1989; 9:895-907.

52. Schwenke DC, Carew TE. Initiation of atherosclerotic lesions in cholesterol-fed rabbits. II. Selective retention of LDL vs. selective increases in LDL permeability in susceptible sites of arteries. Arteriosclerosis 1989; 9:908-918.

53. Steinbrecher UP, Zhang H, Lougheed M. Role of oxidatively modified LDL in atherosclerosis. Free Rad Biol Med 1990; 9:155-68.

54. Walker LN, Reidy MA, Bowyer DE. Morphology and cell kinetics

of fatty streak lesion formation in the hypercholesterolemic rabbit. Am J Pathol 1986; 125:450-9.

55. Schwartz CJ, Valente AJ, Sprague EA et al. The pathogenesis of atherosclerosis: An overview. Clin Cardiol 1991; 14 (2 Suppl 1):I1-I16.

56. Steinberg D, Parthasarathy S, Carew TE et al. Beyond cholesterol: Modifications of low density lipoprotein that increase its atherogenicity. New Engl J Med 1989; 320:915-24.

57. Witztum JL. Current approaches to drug therapy for the hypercholesterolemic patient. Circulation 1989; 80:1101-14.

58. Alderson LM, Endemann G, Lindsey S et al. LDL enhances monocyte adhesion to endothelial cells in vitro. Am J Pathol 1986; 123:334-42.

59. Carew TE. Role of biologically modified low-density lipoprotein in atherosclerosis. Am J Cardiol 1989; 64:18G-22G.

60. Avogaro P, Bittolo BG, Cazzolato G. Presence of a modified low density lipoprotein in humans. Arteriosclerosis 1988; 8:79-87.

61. Harats D, Ben-Naim M, Dabach Y et al. Effect of vitamin C and E supplementation on susceptibility of plasma lipoproteins to peroxidation induced by acute smoking. Atherosclerosis 1990; 85:47-54.

62. Steinberg D. Antioxidants and atherosclerosis. Circulation 1991; 84:1420-5.

63. Regnström J, Nilsson J, Tornvall P et al. Susceptibility to low density lipoprotein oxidation and coronary atherosclerosis in man. Lancet 1992; 339:1183-6.

64. Palinski W, Rosenfeld ME, Ylä-Herttuala S et al. Low density lipoprotein undergoes oxidative modification in vivo. Proc Natl Acad Sci USA 1989; 86:1372-6.

65. Ylä-Herttuala S, Palinski W, Rosenfeld ME et al. Evidence for the presence of oxidatively modified low density lipoprotein in atherosclerotic lesions of rabbit and man. J Clin Invest 1989; 84:1086-95.

66. Cathcart MK, Chisolm GM, McNally AK et al. Oxidative modification of low density lipoprotein (LDL) by activated human monocytes and the cell lines U937 and HL60. In Vitro Cell & Develop Biol 1988; 24:1001-8.

67. Berliner JA, Territo MC, Sevanian A et al. Minimally modified low density lipoprotein stimulates monocyte endothelial interactions. J Clin Invest 1990; 85:1260-6.

68. Liao F, Berliner JA, Mehrabian M et al. Minimally modified low density lipoprotein is biologically active in vivo in mice. J Clin Invest 1991; 87:2253-7.

69. Rajavashisth TB, Andalibi A, Territo MC et al. Induction of endothelial cell expression of granulocyte and macrophage colony-stimulating factors by modified low-density lipoproteins. Nature (Lond) 1990; 344:254-7.

70. Cushing SD, Berliner JA, Valente AJ et al. Minimally modified

low density lipoprotein induces monocyte chemotactic protein 1 in human endothelial cells and smooth muscle cells. Proc Natl Acad Sci USA 1990; 87:5134-8.

71. Fox PL, Chisholm GM, DiCorleto PE. Lipoprotein-mediated inhibition of endothelial cell production of PDGF like protein depends on free radical lipid peroxidation. J Biol Chem 1987; 262:6046-54.

72. Esterbauer H, Dieber-Rothender M, Waeg G et al. Biochemical, structural and functional properties of oxidized low density lipoproteins. Chem Res Toxicol 1990; 3:77-92.

73. Cathcart MK, Morel DW, Chisholm GM. Monocytes and neutrophils oxidize low density lipoprotein making it cytotoxic. J Leukocyte Biol 1985; 38:341-50.

74. Morel DW, DiCorleto PE, Chisholm GM. Endothelial and smooth muscle cells alter low density lipoprotein in vitro by free radical oxidation. Arteriosclerosis 1984; 4: 357-64.

75. Steinbrecher UP, Parthasarathy S, Leake DS et al. Modification of low density lipoprotein by endothelial cells involves lipid peroxidation and degradation of low density lipoprotein phospholipids. Proc Natl Acad Sci USA 1984; 81:3883-7.

76. Heinecke JW, Rosen H, Chait A. Iron and copper promote modification of LDL by human arterial smooth muscle cells. J Clin Invest 1984; 74:1890-4.

77. Scaccini C, Jialal I. LDL modification by activated polymorphonuclear leukocytes: a cellular model of mild oxidative stress. Free Rad Biol Med 1994; 16:49-5.

78. Hogg N, Darley-Usmar VM, Graham A et al. Peroxynitrite and atherosclerosis. Biochem Soc Trans 1993; 21:358-62.

79. Jessup W, Simpson JA, Dean RT. Does superoxide radical have a role in macrophage-mediated oxidative modification of LDL. Atherosclerosis 1993; 99:107-20.

80. Ylä-Herttuala S, Rosenfeld ME, Parthasarathy S et al. Colocalization of 15-lipoxygenase mRNA and protein with epitopes of oxidized low density lipoprotein in macrophage-rich areas of atherosclerotic lesions. Proc Natl Acad Sci USA 1990; 87:6959-63.

81. Cathcart MA, McNally AK, Chisholm GM. Lipoxygenase-mediated transformation of human low density lipoprotein to an oxidized and cytotoxic complex. J Lipid Res 1991; 32:63-70.

82. McNally AK, Chisholm GM, Morel DW et al. Activated human monocytes oxidize low-density lipoprotein by a lipoxygenase-dependent pathway. J Immunol 1990; 145:254-9.

83. Parthasarathy S, Wieland E, Steinberg D. A role for endothelial cell lipoxygenase in the oxidative modification of low density lipoprotein. Proc Natl Acad Sci USA 1989; 86:1046-50.

84. Rankin SM, Parthasarathy S, Steinberg D. Evidence for a dominant role of lipoxygenase(s) in the oxidation of LDL by mouse peritoneal macrophages. J Lipid Res 1991; 32:449-56.

85. Cathcart MA, McNally AK, Morel DW et al. Superoxide anion

participation in human monocyte-mediated oxidation of low-density lipoprotein to a cytotoxin. J Immunol 1989; 142:1963-9.

86. Heinecke JW, Baker L, Rosen H et al. Superoxide-mediated oxidative modification of low-density lipoprotein by arterial smooth muscle cells. J Clin Invest 1986; 77:757-61.

87. Hiramatsu K, Rosen H, Heinecke JW et al. Superoxide initiates oxidation of low-density lipoprotein by human monocytes. Arteriosclerosis 1987; 7:55-60.

88. Steinbrecher UP. Role of superoxide in endothelial-cell modification of low-density lipoproteins. Biochim Biophys Acta 1988; 959:20-30.

89. Frostegård J, Nilsson J, Haegerstrand A et al. Oxidized low density lipoprotein induces differentiation and adhesion of human monocytes and the monocytic cell line U937. Proc Natl Acad Sci 1990; 87:904-8.

90. Lehr HA, Hübner C, Finckh B et al. Role of leukotrienes in leukocyte adhesion following systemic administration of oxidatively modified human low density lipoprotein in hamsters. J Clin Invest 1991; 88:9-14.

91. Lehr HA, Hübner C, Nolte D et al. Oxidatively modified human low-density lipoprotein stimulates leukocyte adherence to the microvascular endothelium in vivo. Res Exp Med 1991; 191:85-90.

92. Lehr HA, Seemüller J, Hübner C et al. Oxidized LDL-induced leukocyte/endothelium interaction in vivo involves the receptor for platelet-activating factor. Arterioscler Thromb 1993; 13:1013-8.

93. Liao L, Asako H, Kurose I et al. Oxidized lipoproteins elicit leukocyte-endothelial cell adhesion in mesenteric venules. FASEB J 1993; 7:A343.

94. Quinn MT, Parthasarathy S, Fong LG et al. Oxidatively modified low density lipoproteins: a potential role in recruitment and retention of monocyte/macrophages during atherogenesis. Proc Natl Acad Sci 1987; 84:2995-8.

95. Berliner JA, Territo M, Almada L et al. Monocyte chemotactic factor produced by large vessel endothelial cells in vitro. Arteriosclerosis 1986; 6:254-8.

96. Cushing SD, Fogelman AM. Monocytes may amplify their recruitment into inflammatory lesions by inducing monocyte chemotactic protein. Artertioscler Thromb 1992; 12:78-82.

97. Navab M, Imes SS, Hama SY et al. Monocyte transmigration induced by modification of low density lipoprotein in cocultures of human aortic wall cells is due to induction of monocyte chemotactic protein 1 synthesis and is abolished by high density lipoprotein. J Clin Invest 1991; 88:2039-46.

98. Ylä-Herttuala S, Lipton BA, Rosenfeld ME et al. Expression of monocyte chemoattractant protein 1 in macrophage-rich areas of human and rabbit atherosclerotic lesions. Proc Natl Acad Sci USA 1991; 88:5252-6.

99. Terkeltaub R, Banka CL, Solan J et al. Oxidized LDL induces

monocytic cell expression of interleukin-8, a chemokine with T-lymphocyte chemotactic activity. Arterioscler Thromb 1994; 14:47-53.

100. Baggiolini M, Walz A, Kunkel SL. Neutrophil-activating Peptide-1/Interleukin 8, a novel cytokine that activates neutrophils. Am Soc Clin Invest 1989; 84:1045-49.

101. Larsen CG, Anderson AO, Appella E et al. The neutrophil-activating protein (NAP-1) is also chemotactic for T lymphocytes. Science 1989; 243:1464-1466.

102. Quinn MT, Parthasarathy S, Steinberg D. Lysophosphatidylcholine: a chemotactic factor for human monocytes and its potential role in atherogenesis. Proc Natl Acad Sci USA 1988; 85:2805-9.

103. Portman OW, Alexander M. Lysophosphatidylcholine concentration and metabolism in aortic intima plus inner media: effect of nutritionally induced atherosclerosis. J Lipid Res 1969; 10:158-165.

104. Kugiyama K, Kerns SA, Morrisett JD et al. Impairment of endothelial-dependent arterial relaxation by lysolecithin in modified low density lipoproteins. Nature (Lond.) 1990; 344:160-162.

105. Yokoyama M, Hirata K, Miyake R et al. Lysophosphatidylcholine: essential role in the inhibition of endothelial-dependent vasorelaxation by oxidized low density lipoprotein. Biochem Biophys Res Commun 1990; 16:301-8.

106. McMurray HF, Parthasarathy S, Steinberg D. Oxidatively modified low density lipoprotein is a chemoattractant for human T lymphocytes. J Clin Invest 1993; 92:1004-8.

107. Quinn MT, Parthasarathy S, Steinberg D. Endothelial cell-derived chemotactic activity for mouse peritoneal macrophages and the effects of modified forms of low density lipoprotein. Proc Natl Acad Sci USA 1985; 82:5949-53.

108. Lehr HA, Kröber M, Hübner C et al. Stimulation of leukocyte/endothelium interaction by oxidized low-density lipoprotein in hairless mice. Lab Invest 1993; 68:388-95.

109. Kurtel H, Liao L, Grisham MB et al. Mechanisms of oxidized chylomicron-induced leukocyte-endothelial cell adhesion. Am J Physiol 1995; 268:H2175-82.

110. Lehr HA, Kress E, Menger MD et al. Cigarette smoke elicits leukocyte adhesion to endothelium in hamsters: inhibition by CuZn-SOD. Free Rad Biol Med 1993; 14:573-81.

111. Gillespie MN, Owasoyo JO, Kojima S et al. Enhanced chemotaxis and superoxide anion production by polymorphonuclear leukocytes from nicotine-treated and smoke-exposed rats. Toxicology 1987; 45:45-52.

112. Sachinidis A, Locher R, Mengden T et al. Low density lipoprotein elevates intracellular calcium and pH in vascular smooth muscle cells and fibroblasts without mediation of LDL receptors. Biochem Biophys Res Commun 1990; 167:353-9.

113. Lehr HA, Kress E, Menger MD. Involvement of 5-lipoxygenase

products in cigarette smoke-induced leukocyte/endothelium interaction in hamsters. Int J Microcirc: Clin Exp 1993; 12:61-73.

114. De Caterina R, Mazzone A, Giannessi D et al. Leukotriene B_4 production in human atherosclerotic plques. Biomed Biochim Acta 1980; 47:S182-5.

115. Patel KD, Zimmerman GA, Prescott SM et al. Novel leukocyte agonists are released by endothelial cells exposed to peroxide. J Biol Chem 1992; 267:15168-74.

116. Smiley PL, Stremler KE, Prescott SM et al. Oxidatively fragmented phosphatidylcholines activate human neutrophils through the receptor for platelet-activating factor. J Biol Chem 1991; 266:11104-10.

117. Prescott SM, Zimmerman GA, McIntyre TM. Platelet-activating factor. J Biol Chem 1990; 265:17381-4.

118. Zimmerman GA, McIntyre TM, Mehra M et al. Endothelial cell-associated platelet-activating factor: a novel mechanism for signaling intercellular adhesion. J Cell Biol 1990; 110:529-40.

119. Grylewski RJ, Palmer RM, Moncada S. Superoxide anion is involved in the breakdown of endothelium-derived relaxing factor. Nature (Lond.) 1986; 320:454-6.

120. Kurose I, Wolf R, Grisham MB et al. Modulation of ischemia/reperfusion-induced microvascular dysfunction by nitric oxide. Circ Res 1994; 74:376-82.

121. Gaboury J, Woodman RC, Granger DN et al. Nitric oxide prevents leukocyte adherence: role of superoxide. Am J Physiol 1993; 265:H862-7.

122. Kubes P, Kurose I, Granger DN. NO donors prevent integrin-induced leukocyte adhesion but not P-selectin-dependent rolling in postischemic venules. Am J Physiol 1994; 267:H931-7.

123. Kubes P, Granger DN. Nitric oxide modulates microvascular permeability. Am J Physiol 1992; 262:H611-5.

124. Kubes P, Kanwar S, Niu XF et al. Nitric oxide synthesis inhibition induces leukocyte adhesion via superoxide and mast cells. FASEB J 1993; 7:1293-9.

125. Yuan Y, Granger HJ, Zawieja DC et al. Histamine increases venular permeability via a phospholipase C-NO synthase-guanylate cyclase cascade. Am J Physiol 1993; 264:H1734-9.

126. McDermott JR. Studies on the catabolism of N^G-methylarginine, N^G,N'^G-dimethylarginine and N^G,N^G-dimethylarginine in the rabbit. Biochem J 1976; 154:179-84.

127. Kimoto M, Tsuji H, Ogawa T. N^G, N^G-dimethyl-L-arginine, a dominant precursor of endogenous dimethylamine in rats. Amino Acids 1994; 6:273-82.

128. Vallance P, Leone A, Calver A et al. Accumulation of an endogenous inhibitor of nitric oxide synthesis in chronic renal failure. Lancet 1992; 339:572-5.

129. Fickling SA, Nussey SS, Vallance P et al. Synthesis of N^G, N^G dimethylarginine from L-arginine by human endothelial cells. En-

dothelium 1993; l(suppl):S16.

130. Kurose I, Wolf RE, Grisham MB et al. Effects of an endogenous inhibitor of nitric oxide synthesis on postcapillary venules. Am J Physiol 1995; 268:H2224-31.

131. Gardiner SM, Kemp PA, Bennett T et al. Regional and cardiac haemodynamic effects of N^G, N^G, dimethyl-L-arginine and their reversibility by vasodilators in conscious rats. Br J Pharmacol 1993; 110:1457-64.

132. Bazzoni GE, Dejana E, Del Maschio A. Platelet-neutrophil interactions, possible relevance in the pathogenesis of thrombosis and inflammation. Haematologica 1991; 76:491-9.

133. Kurose I, Kubes P, Wolf R et al. Inhibition of nitric oxide production. Mechanisms of vascular albumin leakage. Circ Res 1993; 73:164-71.

134. Lehr HA, Frei B, Arfors KE. Vitamin C prevents cigarette smoke-induced leukocyte aggregation and adhesion to endothelium in vivo. Proc Natl Acad Sci U S A 1994; 91:7688-92.

135. Lehr HA, Olofsson AM, Carew TE et al. P-selectin mediates the interaction of circulating leukocytes with platelets and microvascular endothelium in response to oxidized lipoprotein in vivo. Lab Invest 1994; 71:380-6.

136. Harlan JM, Liu DY. Adhesion: its role in inflammatory disease. New York: Freeman, 1992.

137. Golino P, Maroko PR, Carew TE. Efficacy of platelet depletion in counteracting the detrimental effect of acute hypercholesterolemia on infarct size and the no-reflow phenomenon in rabbits undergoing coronary artery occlusion-reperfusion. Circulation 1987; 76:173-80.

A THERAPEUTIC APPLICATION OF NITRIC OXIDE: GI-SPARING NSAIDs

Brian K. Reuter and John L. Wallace

INTRODUCTION

Nonsteroidal anti-inflammatory drugs (NSAIDs) are among the most commonly used drugs,[1] being prescribed mainly for their anti-inflammatory and analgesic properties, but also widely used in over-the-counter preparations for their antipyretic effects and, at least in the case of aspirin, for their antithrombotic effects. The major limitation to the long-term use of NSAIDs is their untoward effects on the gastrointestinal tract. All presently marketed NSAIDs have the capacity to cause significant gastrointestinal bleeding, although the relative toxicity varies considerably among this class of compounds and from patient to patient. NSAIDs can also cause acute and chronic gastric ulcers in the stomach and duodenum, as well as causing significant injury in more distal regions of the small intestine. Furthermore, NSAIDs have been reported to be capable of causing a "reactivation" of quiescent inflammatory bowel disease.[2]

The most important of the adverse effects of NSAIDs, in terms of both frequency and clinical impact, is ulceration of the stomach following repeated administration of these drugs. NSAIDs will

Nitric Oxide: A Modulator of Cell-Cell Interactions in the Microcirculation, edited by Paul Kubes. © 1995 R.G. Landes Company.

cause superficial erosions, primarily in the corpus region and deep (i.e., penetrating through the muscularis mucosae) ulcers in the antral region of the stomach. It is the latter that are more clinically significant,[3] since these ulcers are more likely to bleed and perforate than the superficial erosions.

While several strategies have been employed to reduce the ulcerogenic potential of NSAIDs, none have proven to be effective in reducing clinically significant adverse reactions.[4] One of the problems associated with designing NSAIDs with reduced gastrointestinal toxicity is that the mechanism underlying NSAID-induced gastroenteropathy remains poorly understood. The ability of NSAIDs to suppress prostaglandin synthesis appears to be critical in the pathogenesis of gastric ulceration induced by these agents but may not be of central importance in the production of small intestinal injury.[5]

In this chapter, we will review the mechanisms proposed for NSAID-induced gastropathy and the role of nitric oxide (NO) in mucosal defense. The chapter will also describe a novel series of NSAIDs, modified to incorporate a nitric oxide-releasing moiety. These derivatives exhibit markedly reduced gastrointestinal toxicity while retaining the beneficial effects of the parent compounds.

PATHOGENESIS OF NSAID-INDUCED GASTROPATHY

The mechanism responsible for the ulcerogenic effects of NSAIDs in the stomach has been the subject of considerable debate since the discovery by Vane[6] that NSAIDs suppress prostaglandin synthesis. It is now well known that all NSAIDs have the capacity to inhibit the activity of the enzyme cyclo-oxygenase. Aspirin inhibits this enzyme irreversibly through acetylation of a serine residue at the active site, while the other commonly used NSAIDs (e.g., indomethacin, ibuprofen, naproxen) inhibit the enzyme in a reversible manner. The debate regarding the primary mechanism for NSAID-induced ulceration has focused primarily on the relative contribution to their ulcerogenic actions of suppression of prostaglandin synthesis versus their topical irritant properties. While topical irritant properties likely contribute significantly to the superficial erosions induced by NSAIDs, there is now overwhelming evidence that it is the suppression of gastric prostaglandin synthesis that is responsible for the clinically significant gastric injury.

Why suppression of gastric prostaglandin synthesis leads to ulcer formation is still not well understood.

Studies performed in the early 1990s have provided compelling evidence for a role for neutrophils in the pathogenesis of experimental NSAID-gastropathy (recently reviewed by Wallace[7]). The evidence includes: 1) experimental NSAID-gastropathy is markedly reduced in rats rendered neutropenic through pretreatment with either antineutrophil serum or methotrexate,[8,9] 2) treatment with monoclonal antibodies which prevent leukocyte adherence to the vascular endothelium results in a significant attenuation in the severity of experimental NSAID-gastropathy,[10,11] 3) NSAID administration results in significant increases in the numbers of leukocytes (primarily neutrophils) adhering to the vascular endothelium in the gastric microcirculation and in postcapillary mesenteric venules,[11-15] 4) cytoprotective prostaglandins can prevent NSAID-induced leukocyte aherence,[13,14] and 5) NSAID administration results in a rapid upregulation of the expression of ICAM-1, an important endothelial adhesion molecule, on the microvasculature of the rat gastric mucosa.[16]

NSAID-induced leukocyte adherence could contribute to gastric mucosal injury in two major ways.[17] Firstly, adherence of neutrophils to the vascular endothelium would likely be accompanied by activation of these cells, leading to the liberation of oxygen-derived free radicals and proteases. These substances could mediate much of the endothelial and epithelial injury caused by NSAIDs. Indeed, there is evidence for a role for reactive oxygen metabolites in experimental NSAID-gastropathy.[18] Secondly, neutrophil adherence to the vascular endothelium could lead to capillary obstruction, resulting in a reduction in gastric mucosal blood flow, and thereby predisposing the mucosa to injury. A reduction in gastric blood flow following NSAID administration has been reported by many groups[12,19,20] and has been shown to occur subsequent to the appearance of "white thrombi" in the gastric microcirculation.[12]

Recent evidence implicates tumor necrosis factor (TNF) in the pathogenesis of NSAID-induced gastric damage and leukocyte adherence. TNF is a potent stimulus for ICAM-1 expression on the vascular endothelium, and its release from macrophages can be stimulated by NSAIDs and inhibited by prostaglandins.[21] Santucci et al[22] reported a marked elevation of serum TNF levels three hours after administration of indomethacin to rats. Significant leukocyte margin-ation within the gastric microcirculation was also observed at this time. When they pretreated the rats with pentoxifylline, a dose-dependent reduction of serum TNF levels was observed, as well as parallel re-ductions in leukocyte margination and gastric damage. These studies have been reproduced in our laboratory, and the similar effects to those observed with pentoxifylline have been observed with other TNF synthesis inhibitors.[23]

While TNF is one possible mediator of leukocyte adherence following NSAID administration, it is also possible that these ad-hesive interactions between neutrophils and the endothelium are in part attributable to the suppression by NSAIDs of endothelial production of prostacyclin, a potent inhibitor of neutrophil acti-vation and adherence. Another endothelial-derived mediator which is capable of inhibiting neutrophil function is nitric oxide. Inter-estingly, suppression of nitric oxide administration results in leu-kocyte adherence to the vascular endothelium similar to that ob-served following NSAID administration.[24] Moreover, suppression of nitric oxide synthesis results in a marked exacerbation of the severity of NSAID-induced gastric damage.[25] On the other hand, nitric oxide donors can reduce the severity of NSAID-induced gastric damage.[26]

NO-RELEASING NSAIDS: GI-SPARING EFFECTS

Based on the observations that NO donors could reduce the severity of NSAID-induced gastric damage, while suppressors of NO synthesis augmented such injury, we postulated that addition of a nitric oxide-releasing moiety to NSAIDs would greatly reduce their toxicity in the gastrointestinal tract. In theory, the nitric oxide released from these compounds would maintain mucosal blood flow in the stomach and intestine and would suppress leukocyte activa-tion that normally occurs following NSAID administration. Of course, the utility of such an NSAID derivative would depend on the derivitizaiton not adversely affecting the ability of the NSAID to sup-press cyclo-oxygenase activity, as this seems to be the key action accounting for the anti-inflammatory, antipyretic, analgesic, and antithrombotic effects of NSAIDs. Examples of some of the NO-releasing NSAIDs that we have examined are shown in Figure 7.1.

The first series of experiments we performed with these com-pounds was aimed at testing the hypothesis that the compounds

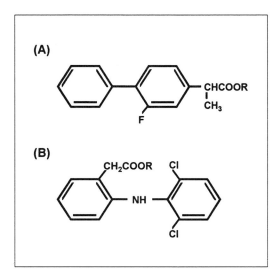

Fig. 7.1. Structures of (A) flurbiprofen (R = H) and (B) diclofenac (R = Na), two common nonsteroidal anti-inflammatory drugs. Their NO-releasing derivatives, flurbiprofen 4-nitroxy-butylester and nitrofenac, are produced by addition of a nitroxybutyl moiety [R = (CH₂)₄-ONO₂] to flurbiprofen and diclofenac, respectively.

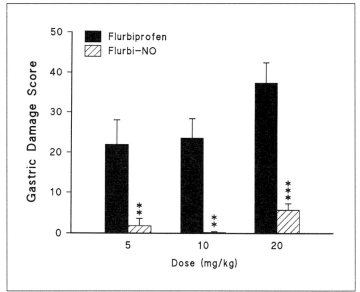

Fig. 7.2. Acute gastric mucosal injury in the rat following the oral administration of either flurbiprofen (5-20 mg/kg) or flurbiprofen-NO (5-20 mg/kg). Gastric damage was assessed 5 hours after drug administration.

would have reduced ulcerogenic activity in the stomach, but would retain the ability to suppress gastric prostaglandin synthesis. As shown in Figure 7.2, flurbiprofen caused significant acute gastric injury within 5 hours of its administration.[27] On the other hand, flurbiprofen-4-nitroxybutylester (flurbi-NO) produced markedly less gastric damage, despite suppressing gastric prostaglandin synthesis as effectively as flurbiprofen (> 80% at all doses tested). Similar observations were made when the nitric oxide-releasing derivatives of ketoprofen and diclofenac were compared to the native NSAIDs,[27,28] and we have unpublished data showing that derivatives of both naproxen and ketorolac have markedly reduced acute gastric toxicity.

The acute gastric injury produced in the rat by NSAIDs occurs primarily in the corpus region of the stomach and is superficial in nature. It has been argued that these lesions are not truly representative of the chronic, antral ulcers observed following long-term NSAID ingestion in man. For this reason, we established a model of chronic antral ulcer induced by NSAIDs in the rabbit.[29] Twice-daily administration of an NSAID to rabbits reproducibly causes antral ulceration that penetrates to the depth of the external muscle layer of the stomach and sometimes perforates. We compared the ability of diclofenac versus its nitric oxide-releasing derivative, nitrofenac, to produce these ulcers. While diclofenac (10 mg/kg) caused antral ulceration in approximately 80% of the rabbits, nitrofenac did not produce any detectable gastric damage.[28]

In addition to causing significant damage to the stomach, NSAIDs are known to cause intestinal damage in many patients. Indeed, it has been argued that the small intestine is the major site of bleeding in patients taking NSAIDs.[30] The pathogenesis of NSAID-enteropathy is not well understood, although there is evidence suggesting that suppression of cyclo-oxygenase activity may not be as critical as is the case for NSAID-gastropathy, while roles for bile, enteric bacteria, and enterohepatic circulation of the NSAID have all been supported by experimental evidence.[31-34] While neutrophils have been implicated in the pathogenesis of NSAID-induced small intestinal injury, this role seems to be more secondary than the role played by these cells in NSAID-induced gastric injury.[35,36]

The ability of NO-NSAIDs to cause intestinal injury was examined in healthy rats and in rats with pre-existing colitis. The latter studies were performed because we have previously found that rats with colitis are much more sensitive to the intesinal-damaging actions of NSAIDs than are healthy rats. In both groups of animals, the NO-NSAID nitrofenac was found to cause little or no small intestinal damage, while diclofenac caused extensive intestinal injury often leading to perforation and death.

The mechanism(s) responsible for gastrointestinal-sparing properties of NO-NSAIDs are presumably related to the generation of nitric oxide by these compounds. We have observed, for example, that while NSAIDs such as diclofenac cause a decrease in gastric mucosal blood flow (Fig. 7.3), the NO-NSAID derivative,

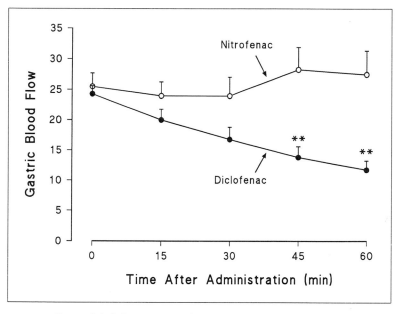

Fig. 7.3. Effects of diclofenac (10 mg/kg) or nitrofenac (15 mg/kg) on rat gastric mucosal blood flow. Blood flow was monitored by laser-Doppler flowmetry for one hour after NSAID or NO-NSAID administration.

nitrofenac, did not produce this effect. Decreased gastric mucosal blood flow is a well characterized consequence of NSAID administration and is believed to conribute significantly to the increased susceptibility of the gastric mucosa to injury following NSAID administration.[37] The NO-NSAIDs also exert effects on leukocyte adherence to the vascular endothelium which are consistent with the release of nitric oxide from these compounds. As mentioned above, NSAIDs stimulate leukocyte adherence to the vascular endothelium. NO-NSAIDs, on the other hand, do not (Fig. 7.4). Both the decrease in gastric mucosal blood flow and adherence of leukocytes to the endothelium following NSAID administration are thought to be, at least in part, attributable to suppression of prostaglandin synthesis. It is important to bear in mind, therefore, that the absence of these effects with NO-NSAIDs occurs despite the fact that these compounds suppress prostaglandin synthesis as effectively as the parent NSAIDs.

NO-NSAIDS: DESIRED EFFECTS

NSAIDs are most commonly prescribed because of their ability to reduce inflammation and to relieve pain. As indicated ear-

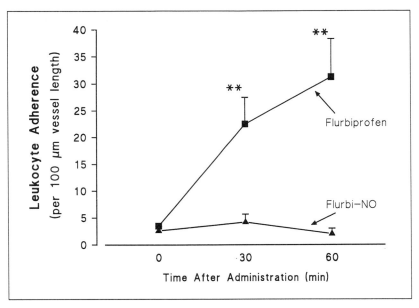

Fig. 7.4. Intravital microscopy determination of leukocyte adherence in rat post-capillary mesenteric venules following oral treatment with flurbiprofen (10 mg/kg) or flurbiprofen-NO (15 mg/kg). Number of adherent leukocytes was determined prior to drug administration and at 30 and 60 minutes thereafter. Reuter & Wallace[12]

lier, the derivatization of the drugs may alter their ability to produce these desired effects. Therefore, NSAIDs and NO-NSAIDs were compared in well characterized animal models of inflammation and pain. Anti-inflammatory activities of diclofenac and nitrofenac were determined in the rat model of carrageenan-induced paw edema.[38] Carrageenan produced a substantial increase in paw volume within hours of its administration in vehicle-pretreated rats. Pretreatment with diclofenac significantly and dose-dependently inhibited the increase in paw volume. Nitrofenac administration also prevented the increase in paw volume in a dose-dependent manner.[28] The NO-NSAID had comparable anti-inflammatory activity to its parent compound. The analgesic properties for another NSAID, ketorolac, and its NO-releasing derivative were examined in the mouse using the phenylquinone (PPQ) writhing assay. The compounds were found to have comparable analgesic activities (unpublished data).

NSAIDs are also used for their antipyretic and antithrombotic effects. Utilizing a rat model of endotoxin-induced fever,[39] the antipyretic activities of diclofenac and nitrofenac were compared. Intraperitoneal injection of endotoxin increased the body temperature of vehicle-treated rats. However, administration of either diclofenac or nitrofenac rapidly suppressed the endotoxin-induced fever, with no significant differences being found between the two compounds.[40] The antithrombotic activities of diclofenac and nitrofenac were examined in both in vitro and in vivo systems. In vitro studies were conducted using human platelets stimulated with thrombin. Preincubation with diclofenac dose-dependently inhibited the aggregation of platelets. Comparable results were obtained when platelets were incubated with nitrofenac. In vivo platelet aggregation was determined following the methods of Pinon.[41] Briefly, rats were orally pretreated with either vehicle, diclofenac or nitrofenac. One hour later the animals received collagen intravenously. Blood samples were obtained three minutes later and the number of platelet aggregates was determined by light microscopy. Both diclofenac and nitrofenac decreased the percentage of platelet aggregates in a dose-dependent manner. At the higher doses tested, the NO-NSAID was found to have enhanced antithrombotic activity compared to diclofenac,[40] presumably attributable to the antiplatelet effects of nitric oxide released from nitrofenac.

These observations that NO-NSAIDs exhibit comparable anti-inflammatory, analgesic, antipyretic and antithrombotic activities to the native NSAID is not surprising, if one considers that the ability of the NSAID to suppress prostaglandin synthesis is unaltered by the derivatization.

SUMMARY

NSAIDs produce their desirable effects by inhibiting the production of prostaglandins. However, the inhibition of prostaglandin synthesis also results in ulceration of the stomach and intestine. This limits the use of NSAIDs and poses problems for the development of new NSAIDs because the beneficial and untoward effects of the drugs are causally linked. Recent evidence that the neutrophil also plays a role in NSAID-gastropathy and that inhibitors of neutrophil activation, prostaglandins and nitric oxide, attenuate the gastric injury associated with NSAID administration led to the development of a novel class of NSAID derivatives. These NO-NSAIDs are composed of a nitric oxide-releasing moiety attached to a standard NSAID. Theoretically, the NSAID portion of the drug should still inhibit prostaglandin synthesis and therefore produce the desirable effects of NSAIDs, while the nitric oxide-generating moiety will prevent the untoward effects by attenuating neutrophil activation. In this chapter we described the effects of NO-NSAIDs on the gastrointestinal tract and determined their therapeutic activity. The NO-NSAIDs were found to produce markedly less ulceration in both acute and chronic models of gastric injury. They were also found to have decreased toxicity in the small intestine and in a model of pre-existing intestinal inflammation. The NO-NSAIDs inhibit prostaglandin synthesis comparable to native NSAIDs, and therefore, retain the desirable effects of these drugs (i.e., anti-inflammatory, analgesic, antipyretic and antithrombotic). Therefore, NO-NSAIDs may offer an alternative to existing NSAIDs with markedly reduced GI side effects.

ACKNOWLEDGMENTS

This work was supported by grants from the Medical Research Council (MRC) of Canada. Dr. Wallace is an MRC Scientist and an Alberta Heritage Foundation for Medical Research Scientist.

REFERENCES

1. Garner A. Adaptation in the pharmaceutical industry, with particular reference to gastrointestinal drugs and diseases. Scand J Gastroenterol 1992; 27 (suppl. 193):83-89.
2. Kaufmann HJ, Taubin HL. Nonsteroidal anti-inflammatory drugs activate quiescent inflammatory bowel disease. Ann Intern Med 1987; 107:513-516.
3. McCarthy DM. NSAID-induced gastrointestinal damage—a critical review of prophylaxis and therapy. J Clin Gastroenterol 1990; 12 (suppl. 2):S13-S20.
4. Soll AH, Weinstein WM, Kurata J et al. Nonsteroidal anti-inflammatory drugs and peptic ulcer disease. Ann Intern Med 1991; 114: 307-319.
5. Whittle BJR. Temporal relationship between cyclooxygenase inhibition, as measured by prostacyclin biosynthesis, and the gastrointestinal damage induced by indomethacin in the rat. Gastroenterology 1981; 80:94-8.
6. Vane JR. Inhibition of prostaglandin synthesis as a mechanism of action for aspirin-like drugs. Nature New Biol 1971; 231: 232-235.
7. Wallace JL. Gastric ulceration: critical events at the neutrophil-endothelium interface. Can J Physiol Pharmaol 1993; 71:98-102.
8. Wallace JL, Keenan CM, Granger DN. Gastric ulceration induced by nonsteroidal anti-inflammatory drugs is a neutrophil-dependent process. Am J Physiol 1990; 259:G462-G467.
9. Lee M, Lee AK, Feldman M. Aspirin-induced acute gastric mucodal injury is a neutrophil-dependent process in rats. Am J Physiol 1992;

263:G920-G926.

10. Wallace JL, Arfors K-E, McKnight GW. A monoclonal antibody against the CD18 leukocyte adhesion molecule prevents indomethacin-induced gastric damage in the rabbit. Gastroenterology 1991; 100:878-883.

11. Wallace JL, McKnight W, Miyasaka M et al. Role of endothelial adhesion molecules in NSAID-induced gastric mucosal injury. Am J Physiol 1993a; 265:G993-G998.

12. Kitahora T, Guth PH. Effect of aspirin plus hydrochloric acid on the gastric mucosal microcirculation. Gastroenterology 1987; 93:810-817.

13. Asako H, Kubes P, Wallace JL et al. Indomethacin-induced leukocyte adhesion in mesenteric venules: role of lipoxygenase products. Am J Physiol 1992a; 262:G903-G908.

14. Asako H, Kubes P, Wallace JL et al. Modulation of leukocyte adhesion to rat mesenteric venules by aspirin and salicylate. Gastroenterology 1992b; 103:146-152.

15. Wallace JL, McCafferty D-M, Carter L et al. Tissue-selective inhibition of prostaglandin synthesis in rat by tepoxalin: anti-inflammatory without gastropathy. Gastroenterology 1993b; 105:1630-1636.

16. Andrews FJ, Malcontenti-Wilson C, O'Brien PE. Effect of nonsteroidal anti-inflammatory drugs on LFA-1 and ICAM-1 expression in gastric mucosa. Am J Physiol 1994; 266:G657-G664.

17. Wallace JL, Granger DN. The pathogenesis of NSAID-gastropathy—are neutrophils the culprits? Trends Pharmacol Sci 1992; 13:129-131.

18. Vaananen PM, Meddings JB, Wallace JL. Role of oxygen-derived free radicals in indomethacin-induced gastric injury. Am J Physiol 1991; 261:G470-G475.

19. Ashley SW, Sonnenschein LA, Cheung LY. Focal gastric mucosal blood flow at the site of aspirin-induced ulceration. Am J Surg 1985; 149:53-59.

20. Gana TJ, Huhlewych R, Koo J. Focal gastric mucosal blood flow in aspirin-induced ulceration. Ann Surg 1987; 205:399-403.

21. Kunkel SL, Wiggins RC, Chensue SW et al. Regulation of macrophage tumor necrosis factor production by prostaglandin E2. Biochem Biophys Res Commun 1986; 137:404-410.

22. Santucci L, Fiorucci S, Giansanti M et al. Pentoxifylline prevents indomethacin induced acute gastric mucosal damage in rats: role of tumor necrosis factor alpha. Gut 1994; 35:909-915.

23. Appleyard CB, Tigley AW, Swain MG et al. TNFα is a critical mediator of NSAID-induced gastric ulceration in rats. Gastroenterology 1995; 108:A47 (abstract).

24. Kubes P, Suzuki M, Granger DN. Nitric oxide: an endogenous modulator of leukocyte adhesion. Proc Natl Acad Sci USA 1991; 88:4651-4655.

25. Whittle BJR. Neuronal and endothelium-derived mediators in the modulation of the gastric microcirculation: integrity in the balance. Br J Pharmacol 1993; 110:3-17.

26. Wallace JL, Reuter BK, Cirino G. Nitric oxide-releasing NSAIDs: a novel approach for reducing gastrointestinal toxicity. J Gastroenterol Hepatol 1994a; 9 (suppl. 1):S40-S44.

27. Wallace JL, Reuter B, Cicala C et al. Novel nonsteroidal anti-inflammatory drug derivatives with markedly reduced ulcerogenic properties in the rat. Gastroenterology 1994b; 107:173-179.

28. Wallace JL, Reuter B, Cicala C et al. A diclofenac derivative without ulcerogenic properties. Eur J Pharmacol 1994c; 257:249-255.

29. Wallace JL, McKnight GW. Characterization of a simple animal model for nonsteroidal anti-inflammatory drug induced antral ulcer. Can J Physiol Pharmacol 1993; 71:447-452.

30. Bjarnason I, Hayllar J, MacPherson AJ et al. Side effects of nonsteroidal anti-inflammatory drugs on the small and large intestine in humans. Gastroenterology 1993; 104:1832-1847.

31. Yamada T, Deitch E, Specian RD et al. Mechanisms of acute and chronic intestinal inflammation induced by indomethacin. Inflammation 1993; 17:641-662.

32. Robert A, Asano T. Resistance of germfree rats to indomethain-induced intestinal lesions. Prostaglandins 1977; 14:333-341.

33. Kent TH, Cardelli RM, Stamler FW. Small intestinal ulcers and intestinal flora in rats given indomethacin. Am J Pathol 1969; 54:237-249.

34. Wax J, Clinger WA, Varner P et al. Relationship of the enterohepatic cycle to the ulcerogenesis in the rat small bowel with flufenamic acid. Gastroenterology 1970; 58:772-780.

35. Miura S, Suematsu M, Tanaka S et al. Microcirculatory disturbance in indomethacin-induced intestinal ulcer. Am J Physiol 1991; 261:G213-G219.

36. Bjarnason I, Hayllar J, Smethurst P et al. Metronidazole reduces intestinal inflammation and blood loss in non-steroidal anti-inflammatory drug induced enteropathy. Gut 1992; 33:1204-1208.

37. Wallace JL, McKnight GW. The mucoid cap over superficial gastric damage in the rat. A high-pH environment dissipated by nonsteroidal anti-inflammatory drugs and endothelin. Gastroenterology 1990; 99:295-304.

38. Cirino G, Peers SH, Floer RJ et al. Human recombinant lipocortin 1 has acute local anti-inflammatory properties in the rat paw edema test. Proc Natl Acad Sci USA 1989; 86:3428-3432.

39. Monda M, Pittman QJ. Cortical spreading depression blocks prostaglandin E1 and endotoxin fever in rats. Am J Physiol 1993; 264:R456-R459.

40. Wallace JL, Pittman QJ, Cirino G. Nitric oxide releasing NSAIDs: a novel class of GI-sparing anti-inflammatory drugs. In: Proznansky W, ed. New Molecular Approaches to Anti-Inflammatory Therapy. Basel: Birkhauser Press, 1995; 121-129.

41. Pinon JF. In vivo study of platelet aggregation in the rat. J Pharmacol Methods 1984; 12:79-84.

INDEX

Page numbers in italics denote figures (f) and tables (t).

MEDICAL INTELLIGENCE UNIT

AVAILABLE AND UPCOMING TITLES

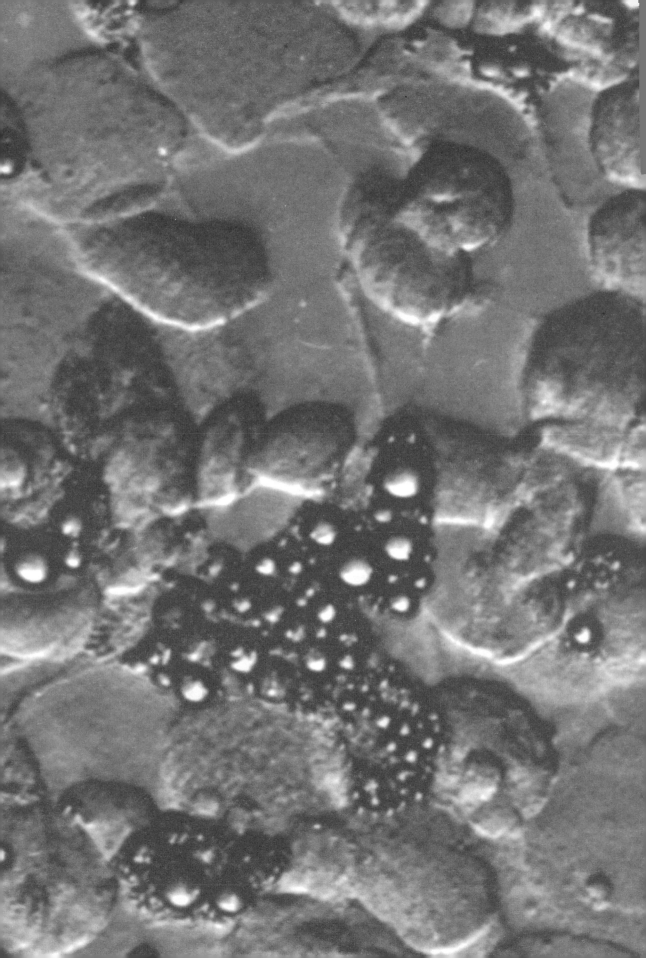